MCAT® Behavioral Sciences

2025–2026 Edition: An Illustrated Guide

Copyright © 2024
On behalf of UWorld, LLC
Dallas, TX
USA

All rights reserved.
Printed in English, in the United States of America.

Reproduction or translation of any part of this work beyond that permitted by Sections 107 and 108 of the United States Copyright Act without the permission of the copyright owner is unlawful.

The Medical College Admission Test (MCAT®) and the United States Medical Licensing Examination (USMLE®) are registered trademarks of the Association of American Medical Colleges (AAMC®). The AAMC® neither sponsors nor endorses this UWorld product.

Facebook® and Instagram® are registered trademarks of Facebook, Inc. which neither sponsors nor endorses this UWorld product.

X is an unregistered mark used by X Corp, which neither sponsors nor endorses this UWorld product.

Acknowledgments for the 2025–2026 Edition

Ensuring that the course materials in this book are accurate and up to date would not have been possible without the multifaceted contributions from our team of content experts, editors, illustrators, software developers, and other amazing support staff. UWorld's passion for education continues to be the driving force behind all our products, along with our focus on quality and dedication to student success.

About the MCAT Exam

Taking the MCAT is a significant milestone on your path to a rewarding career in medicine. Scan the QR codes below to learn crucial information about this exam as you take your next step before medical school.

Basic MCAT Exam Information

Scores and Percentiles

MCAT Sections

Registration Guide

Preparing for the MCAT with UWorld

The MCAT is a grueling exam spanning seven subjects that is designed to test your aptitude in areas essential for success in medicine. Preparing for the exam can be intimidating—so much so that in post-MCAT questionnaires conducted by the AAMC®, a majority of students report not feeling confident about their MCAT performance.

In response, UWorld set out to create premier learning tools to teach students the entire MCAT syllabus, both efficiently and effectively. Taking what we learned from helping over 90% of medical students prepare for their medical board exams (USMLE®), we launched the UWorld MCAT Qbank in 2017 and the UWorld MCAT UBooks in 2024. The MCAT UBooks are meticulously written and designed to provide you with the knowledge and strategies you need to meet your MCAT goals with confidence and to secure your future in medical school.

Below, we explain how to use the MCAT UBooks and MCAT Qbank together for a streamlined learning experience. By strategically integrating both resources into your study plan, you will improve your understanding of key MCAT content as well as build critical reasoning skills, giving you the best chance at achieving your target score.

MCAT UBooks: Illustrated and Annotated Guides

The MCAT UBooks include not only the printed editions for each MCAT subject but also provide digital access to interactive versions of the same books. There are eight printed MCAT UBooks in all, six comprehensive review books covering the science subjects and two specialized books for the Critical Analysis and Reasoning Skills (CARS) section of the exam:

- Biology
- Biochemistry
- General Chemistry
- Organic Chemistry
- Physics
- Behavioral Sciences
- CARS (Annotated Practice Book)
- CARS Passage Booklet (Annotated)

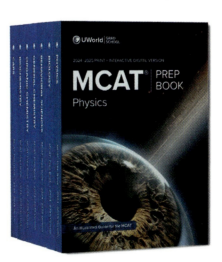

Each UBook is organized into Units, which are divided into Chapters. The Chapters are then split into Lessons, which are further subdivided into Concepts.

MCAT Sciences: Printed UBook Features

The MCAT UBooks bring difficult science concepts to life with thousands of engaging, high-impact visual aids that make topics easier to understand and retain. In addition, the printed UBooks present key terms in blue, indicating clickable illustration hyperlinks in the digital version that will help you learn more about a scientific concept.

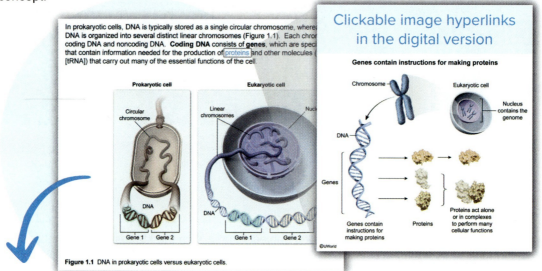

Thousands of educational illustrations in the print book

Clickable image hyperlinks in the digital version

Test Your Basic Science Knowledge with Concept Check Questions

The printed UBooks also include 450 new questions—never before available in the UWorld Qbank—for Biology, General Chemistry, Organic Chemistry, Biochemistry, and Physics. These new questions, called Concept Checks, are interspersed throughout the entire book to enhance your learning experience. Concept Checks allow you to instantly test yourself on MCAT concepts you just learned from the UBook.

Short answers to the Concept Checks are found in the appendix at the end of each printed UBook. In addition, the digital version of the UBook provides an interactive learning experience by giving more detailed, illustrated, step-by-step explanations of each Concept Check. These enhanced explanations will help reinforce your learning and clarify any areas of uncertainty you may have.

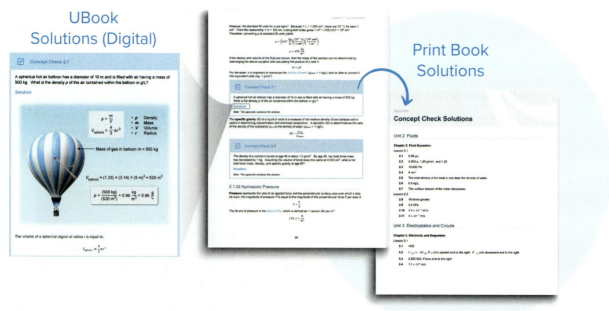

UBook Solutions (Digital)

Print Book Solutions

MCAT CARS Printed UBook Features

For CARS, the main book, or Annotated Practice Book, teaches you the specialized CARS skills and strategies you need to master and then follows up with multiple sets of MCAT-level practice questions.

Additionally, the CARS Passage Booklet includes annotated versions of the passages in the CARS Main Book. From these annotations, you will learn how to break down a CARS passage in a step-by-step manner to find the right answer to each CARS question.

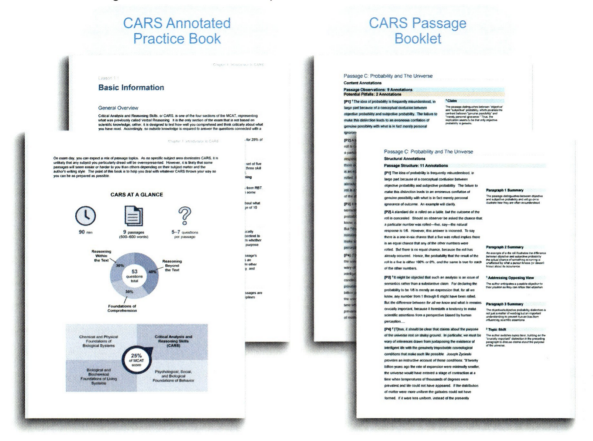

MCAT-Level Exam Practice with the UWorld Qbank

UWorld's MCAT UBooks and Qbank were designed to be used together for a comprehensive review experience. The UWorld Qbank provides an active learning approach to MCAT prep, with thousands of MCAT-level questions that align with each UBook.

The printed UBooks include a prompt at the end of each unit that explains how to access unit practice tests in the MCAT Qbank. In addition, the MCAT UBooks' digital platform enables you to easily create your own unit tests based on each MCAT subject.

To purchase MCAT Qbank access or to begin a free seven-day trial, visit gradschool.uworld.com/mcat.

Boost Your Score with the #1 MCAT Qbank

Why use the UWorld Qbank?

- Thousands of high-yield MCAT-level questions
- In-depth, visually engaging answer explanations
- Confidence-building user interface identical to the exam
- Data-driven performance and improvement tracking
- Fully featured mobile app for on-the-go review

Special Features Integrating Digital UBooks and the UWorld Qbank

The digital MCAT UBooks and the MCAT Qbank come with several integrated features that transform ordinary reading into an interactive study session. These time-saving tools enable you to personalize your MCAT test prep, get the most out of our detailed explanations, save valuable time, and know when you are ready for exam day.

My Notebook

My Notebook, a personalized note-taking tool, allows you to easily copy and organize content from the UBooks and the Qbank. Simplify your study routine by efficiently recording the MCAT content you will encounter in the exam, and streamline your review process by seamlessly retrieving high-yield concepts to boost your study performance—in less time.

Digital Flashcards

Our unique flashcard feature makes it easy for students to copy definitions and images from the MCAT UBooks and Qbank into digital flashcards. Each card makes use of spaced repetition, a research-supported learning methodology that improves information retention and recall. Based on how you rate your understanding of flashcard content, our algorithm will display the card more or less frequently.

Fully Featured Mobile App

Study for your MCAT exams anytime, anywhere, with our industry-leading mobile app that provides complete access to your MCAT prep materials and that syncs seamlessly across all devices. With the UWorld MCAT app, you can catch up on reading, flip through flashcards between classes, or take a practice quiz during lunch to make the most of your downtime and keep MCAT material top of mind.

Book and Qbank Progress Tracking

Track your progress while using the MCAT UBooks and Qbank, and review MCAT content at your own pace. Our learning tools are enhanced by advanced performance analytics that allow users to assess their preparedness over time. Hone in on specific subjects, foundations, and skills to iron out any weaknesses, and even compare your results with those of your peers.

Table of Contents

UNIT 1 THEORIES AND RESEARCH IN PSYCHOLOGY

CHAPTER 1 THEORETICAL APPROACHES IN PSYCHOLOGY 1
- Lesson 1.1 The Psychoanalytic Perspective 3
- Lesson 1.2 The Behaviorist Perspective 4
- Lesson 1.3 The Humanistic Perspective 5

CHAPTER 2 RESEARCH IN PSYCHOLOGY 6
- Lesson 2.1 The Scientific Method 6
- Lesson 2.2 Types of Studies 7
- Lesson 2.3 Results and Conclusions 8
- Lesson 2.4 Ethical Considerations 9

CHAPTER 3 STATISTICS IN PSYCHOLOGY 10
- Lesson 3.1 Descriptive Statistics 10
- Lesson 3.2 Inferential Statistics 11

UNIT 2 BIOLOGICAL BASIS OF BEHAVIOR

CHAPTER 4 THE NERVOUS AND ENDOCRINE SYSTEMS 13
- Lesson 4.1 The Nervous System 15
- Lesson 4.2 Neurons and Neural Communication 19
- Lesson 4.3 The Brain 29
- Lesson 4.4 The Spinal Cord 34
- Lesson 4.5 The Endocrine System 36

CHAPTER 5 HEREDITY AND THE ENVIRONMENT 39
- Lesson 5.1 Behavioral Genetics 39

CHAPTER 6 HUMAN PHYSIOLOGICAL DEVELOPMENT 41
- Lesson 6.1 Human Physiological Development 41

CHAPTER 7 STUDYING THE BRAIN 42
- Lesson 7.1 Neuroimaging Techniques 42
- Lesson 7.2 Other Methods Used in Studying the Brain 43

UNIT 3 SENSATION, PERCEPTION, AND CONSCIOUSNESS

CHAPTER 8 SENSATION 45
- Lesson 8.1 Principles of Sensation 47
- Lesson 8.2 Sensory Receptors 50

CHAPTER 9 PERCEPTION 52
- Lesson 9.1 Principles of Perception 52

CHAPTER 10 VISION 57
- Lesson 10.1 Eye Structure and Function 57
- Lesson 10.2 Visual Processing 59

CHAPTER 11 HEARING 61
- Lesson 11.1 Ear Structure and Function 61
- Lesson 11.2 Auditory Processing 63

CHAPTER 12 OTHER SENSES 65
- Lesson 12.1 Somatosensation 65
- Lesson 12.2 Taste 67
- Lesson 12.3 Smell 69
- Lesson 12.4 The Kinesthetic Sense 71
- Lesson 12.5 The Vestibular Sense 73

CHAPTER 13 CONSCIOUSNESS AND SLEEP 75
Lesson 13.1 States of Consciousness 75
Lesson 13.2 The Stages of Sleep 77

CHAPTER 14 CONSCIOUSNESS-ALTERING SUBSTANCES 81
Lesson 14.1 Consciousness-Altering Substances 81
Lesson 14.2 Problematic Substance Use 82

UNIT 4 LEARNING, MEMORY, AND COGNITION

CHAPTER 15 ATTENTION 85
Lesson 15.1 Selective and Divided Attention 87

CHAPTER 16 NON-ASSOCIATIVE LEARNING 88
Lesson 16.1 Habituation and Dishabituation, Sensitization and Desensitization 88

CHAPTER 17 ASSOCIATIVE LEARNING 90
Lesson 17.1 Classical Conditioning 90
Lesson 17.2 Operant Conditioning 95
Lesson 17.3 The Cognitive Underpinnings of Associative Learning 100
Lesson 17.4 The Biological Underpinnings of Associative Learning 102

CHAPTER 18 OBSERVATIONAL LEARNING 104
Lesson 18.1 The Process of Observational Learning 104
Lesson 18.2 The Biological Underpinnings of Observational Learning 105

CHAPTER 19 MEMORY 106
Lesson 19.1 Encoding, Storage, and Retrieval 106
Lesson 19.2 Forgetting 113
Lesson 19.3 The Biological Underpinnings of Memory 116

CHAPTER 20 COGNITION 119
Lesson 20.1 Cognition Across the Lifespan 119
Lesson 20.2 Theories of Intelligence 123

CHAPTER 21 PROBLEM-SOLVING AND DECISION-MAKING 125
Lesson 21.1 Types of Problem-Solving 125
Lesson 21.2 Barriers to Effective Problem-Solving 126

CHAPTER 22 LANGUAGE 129
Lesson 22.1 Theories of Language Development 129
Lesson 22.2 Language and Cognition 131
Lesson 22.3 The Biological Underpinnings of Language and Speech 132

UNIT 5 MOTIVATION, EMOTION, ATTITUDES, PERSONALITY, AND STRESS

CHAPTER 23 MOTIVATION 135
Lesson 23.1 Influences on Motivation 137
Lesson 23.2 Theories of Motivation 138

CHAPTER 24 EMOTION 140
Lesson 24.1 The Principles and Components of Emotion 140
Lesson 24.2 Theories of Emotion 141
Lesson 24.3 The Biological Underpinnings of Emotion 143

CHAPTER 25 ATTITUDES 145
Lesson 25.1 Attitudes and Behavior 145

CHAPTER 26 PERSONALITY THEORIES 148
Lesson 26.1 Psychoanalytic Theories of Personality 148
Lesson 26.2 Humanistic Theories of Personality 151
Lesson 26.3 Trait Theories of Personality 153

CHAPTER 27 STRESS 155

    Lesson 27.1 The Principles and Components of Stress 155
    Lesson 27.2 The Effects of Stress and Stress Management 157
CHAPTER 28 THEORIES OF ATTITUDE AND BEHAVIOR CHANGE 159
    Lesson 28.1 Theories of Attitude and Behavior Change 159

UNIT 6 PSYCHOLOGICAL DISORDERS AND TREATMENT

CHAPTER 29 PSYCHOLOGICAL DISORDERS 161
    Lesson 29.1 Types of Psychological Disorders 163
    Lesson 29.2 Neurological Disorders 166
CHAPTER 30 THE BIOLOGICAL UNDERPINNINGS OF PSYCHOLOGICAL AND NEUROLOGICAL DISORDERS 167
    Lesson 30.1 The Biological Underpinnings of Psychological Disorders 167
    Lesson 30.2 The Biological Underpinnings of Neurological Disorders 168
CHAPTER 31 TREATMENT OF PSYCHOLOGICAL AND NEUROLOGICAL DISORDERS 169
    Lesson 31.1 Treatment Approaches and Techniques 169

UNIT 7 THEORIES AND RESEARCH IN SOCIOLOGY

CHAPTER 32 THEORETICAL APPROACHES IN SOCIOLOGY 173
    Lesson 32.1 Major Approaches in Sociology 175
    Lesson 32.2 Sociological Theories 176
CHAPTER 33 RESEARCH IN SOCIOLOGY 179
    Lesson 33.1 Empiricism in Sociology 179
    Lesson 33.2 Types of Studies in Sociology 180

UNIT 8 IDENTITY AND SOCIAL INTERACTION

CHAPTER 34 CULTURE 183
    Lesson 34.1 The Components of Culture 187
    Lesson 34.2 Types of Culture 189
    Lesson 34.3 Cultural Change 191
    Lesson 34.4 Socialization 194
CHAPTER 35 IDENTITIES AND IDENTITY FORMATION 196
    Lesson 35.1 Types of Identities 196
    Lesson 35.2 Identity Formation 197
CHAPTER 36 INTERACTING WITH OTHERS 200
    Lesson 36.1 The Presentation of Self 200
    Lesson 36.2 Status 202
    Lesson 36.3 Roles 203
    Lesson 36.4 Groups 205
    Lesson 36.5 Networks 207
    Lesson 36.6 Organizations 209
CHAPTER 37 ATTRACTION, AGGRESSION, AND ATTACHMENT 212
    Lesson 37.1 Attraction 212
    Lesson 37.2 Aggression 213
    Lesson 37.3 Attachment 214
CHAPTER 38 SOCIAL BEHAVIOR IN ANIMALS 215
    Lesson 38.1 Altruism 215
CHAPTER 39 ATTRIBUTING BEHAVIOR TO OTHERS 216
    Lesson 39.1 Attributional Processes 216
CHAPTER 40 PREJUDICE, STEREOTYPES, AND DISCRIMINATION 220
    Lesson 40.1 Prejudice 220
    Lesson 40.2 Stereotypes 221

    Lesson 40.3 Discrimination 223

CHAPTER 41 GROUP PROCESSES AND BEHAVIOR 224
    Lesson 41.1 Group Processes 224
    Lesson 41.2 Social Loafing, the Bystander Effect, and Deindividuation 225
    Lesson 41.3 Conformity and Obedience 227
    Lesson 41.4 Group Decision-Making 229

CHAPTER 42 NORMATIVE AND NON-NORMATIVE BEHAVIOR 231
    Lesson 42.1 Social Norms 231
    Lesson 42.2 Deviance 234

UNIT 9 DEMOGRAPHICS AND SOCIAL STRUCTURE

CHAPTER 43 SOCIAL INSTITUTIONS 237
    Lesson 43.1 Education 239
    Lesson 43.2 Family 241
    Lesson 43.3 Religion 242
    Lesson 43.4 Government and Economy 244
    Lesson 43.5 Medicine 245

CHAPTER 44 DEMOGRAPHIC STRUCTURE OF SOCIETY 247
    Lesson 44.1 Age 247
    Lesson 44.2 Gender 248
    Lesson 44.3 Sexual Orientation 249
    Lesson 44.4 Race and Ethnicity 250
    Lesson 44.5 Immigration Status 251

CHAPTER 45 SOCIAL CLASS AND INEQUALITY 252
    Lesson 45.1 Social Stratification 252
    Lesson 45.2 Social Mobility 256
    Lesson 45.3 Spatial Inequality 258
    Lesson 45.4 Poverty 260
    Lesson 45.5 Health and Healthcare Disparities 262

CHAPTER 46 SOCIAL CHANGE 264
    Lesson 46.1 Urbanization 264
    Lesson 46.2 Globalization 265
    Lesson 46.3 Social Movements 266
    Lesson 46.4 Demographic Change 267

INDEX 271

x

Unit 1 Theories and Research in Psychology

Chapter 1 Theoretical Approaches in Psychology

1.1 The Psychoanalytic Perspective

    1.1.01   The Psychoanalytic Perspective

1.2 The Behaviorist Perspective

    1.2.01   The Behaviorist Perspective

1.3 The Humanistic Perspective

    1.3.01   The Humanistic Perspective

Chapter 2 Research in Psychology

2.1 The Scientific Method

    2.1.01   Hypotheses
    2.1.02   Variables

2.2 Types of Studies

    2.2.01   Correlational Studies
    2.2.02   Experimental Studies

2.3 Results and Conclusions

    2.3.01   Placebos
    2.3.02   Reliability and Validity

2.4 Ethical Considerations

    2.4.01   Informed Consent

Chapter 3 Statistics in Psychology

3.1 Descriptive Statistics

    3.1.01   Descriptive Statistics

3.2 Inferential Statistics

    3.2.01   Correlations

Lesson 1.1
The Psychoanalytic Perspective

1.1.01 The Psychoanalytic Perspective

A number of psychological theories have been developed to explain phenomena related to the mind and behavior.

One of the most influential figures in the history of psychology, Sigmund Freud, was an Austrian physician. Around the turn of the twentieth century, Freud developed the **psychoanalytic approach** (or perspective) in an attempt to treat his patients who sought help for their psychological problems. Through his work with these patients, he advanced the notion of the **unconscious mind**, lying just beyond conscious awareness. Freud's theories focused on how unconscious factors (eg, drives, conflicts stemming from childhood) impact human development and behavior.

Although controversial at times (eg, for its focus on sex and aggression), Freud's psychoanalytic approach continues to influence psychology to this day; many of his followers (ie, "Neo-Freudians" such as Carl Jung and Alfred Adler) developed influential theories of their own. The application of the psychoanalytic (or psychodynamic) perspective to personality is covered in Lesson 26.1.

Lesson 1.2
The Behaviorist Perspective

1.2.01 The Behaviorist Perspective

In contrast to psychoanalytic psychology's focus on the conscious and unconscious aspects of the mind (Lesson 1.1), the **behaviorist perspective** emerged in the early 1900s to emphasize the scientific study of observable actions. In the 1920s, John Watson, considered by many to be the founder of behaviorism, built off the work of Ivan Pavlov (Lesson 17.1) and studied classically conditioned fear. In the 1930s, B.F. Skinner studied the impact of reinforcement and punishment on behavior in his studies on operant conditioning (Lesson 17.2).

These early behaviorists dismissed the study of mental processes to focus solely on overt, observable behavior. Although few psychologists would advocate for this extreme stance today, the behaviorist perspective continues to play a critical role in understanding learning (eg, Chapter 17), personality, and the development and treatment of psychological disorders (Lesson 31.1).

Lesson 1.3
The Humanistic Perspective

1.3.01 The Humanistic Perspective

In the 1960s, the **humanistic perspective** arose in response to the psychoanalytic (Lesson 1.1) and behaviorist (Lesson 1.2) approaches, which were thought to be too pessimistic and mechanistic, respectively. Instead, humanistic psychology takes a more holistic approach to the individual, emphasizing the higher aspects of human nature.

For example, **Abraham Maslow** proposed that humans are motivated to achieve needs in a hierarchy of importance, and at the highest level, self-actualization (ie, fulfilling one's greatest potential) is possible. Another humanistic psychologist, **Carl Rogers**, emphasized the concept of unconditional positive regard (ie, acceptance and support, regardless of someone's behavior). The humanistic perspective continues to serve an important role in psychology, from understanding personality (Lesson 26.2) to understanding and treating psychological disorders (Lesson 31.1).

Lesson 2.1
The Scientific Method

2.1.01 Hypotheses

Psychology is the scientific study of mind and behavior. (Note: this chapter covers research in psychology, although many of the terms are also relevant to sociology. See Chapter 33 for more information on research in sociology.)

A scientific **hypothesis** is a testable explanation for a phenomenon. The **alternative hypothesis** is based on prior evidence and assumes that a significant relationship or difference exists between variables (and often predicts the nature of that relationship or difference).

For example, if a researcher developed a novel intervention for social anxiety (see Concept 29.1.01) and wanted to test if it was more effective in lowering social anxiety than an educational program (see placebo in Concept 2.3.01), the alternative hypothesis could be that the group that received the novel intervention will have significantly lower levels of social anxiety than the group that received the education.

The inverse of the alternative hypothesis, the **null hypothesis**, states that there is no significant difference or relationship between the variables. In this example, the null hypothesis would be that there is no significant difference in levels of social anxiety between the group that received the novel intervention and the group that received the education.

2.1.02 Variables

In an experiment (see Concept 2.2.02), the **independent variable** is manipulated by the experimenter to determine if changes to the independent variable impact the **dependent variable**, which is the outcome that is measured.

Using the example in Concept 2.1.01, if a researcher wanted to test whether participation in a novel intervention for social anxiety lowers levels of social anxiety, the independent variable would be participation in the novel intervention, and the dependent variable would be the level of social anxiety (ie, the measured outcome).

Lesson 2.2
Types of Studies

2.2.01 Correlational Studies

Correlational studies describe relationships between variables but cannot demonstrate cause and effect because no variables are being manipulated. In these research designs, the association between predictor variables and outcome variables may be assessed, although specific causal relationships cannot be determined.

For example, in a study that examines the relationship between the number of close friendships (ie, predictor variable) and the number of serious health problems (ie, outcome variable), researchers would not change the participants' number of close friendships to examine the impact on health. Therefore, this is a correlational research design because it does not involve the manipulation of a variable and cannot demonstrate cause and effect.

2.2.02 Experimental Studies

In contrast to correlational research designs (Concept 2.2.01), in an **experimental study** (also called an experiment), the researcher manipulates the independent variable (Concept 2.1.02) (eg, participation in a novel intervention for social anxiety) to determine its impact on the dependent variable, or outcome (eg, level of social anxiety).

Because changes to the dependent variable are assumed to be caused by the independent variable, experiments can demonstrate a cause-and-effect relationship between variables.

Lesson 2.3
Results and Conclusions

2.3.01 Placebos

A variety of criteria are used to judge the extent to which scientific results can be trusted or believed (ie, are credible). The overall study design is one such criterion that must be examined.

One component used in some experimental designs is a **placebo**, an inactive substance or sham procedure that serves as a comparison to the intervention of interest. Although placebo groups are often helpful in assessing the true benefit of a treatment, a placebo group may be unethical if a sham procedure could result in harm (eg, sham surgery).

The **placebo effect** occurs when the use or application of an inactive substance or sham procedure (ie, a placebo) corresponds with a change in the recipient, often due to their expectations concerning the intervention. For example, a research participant who is given a sugar pill (placebo) instead of a new medicine being tested experiences an improvement of symptoms due to their expectation that the treatment will work.

2.3.02 Reliability and Validity

As Concept 2.3.01 states, scientific findings must be evaluated for their credibility (ie, the extent to which they can be trusted or believed). Both the overall study design and particular measures (eg, surveys) used in a study must be examined for reliability and validity.

Reliability refers to the extent to which an experiment or measure can consistently produce similar results (eg, a test produces a similar score for someone who takes it twice).

In contrast, **validity** refers to the accuracy of a study or measure. **Internal validity** refers to the extent to which a measure or experiment produces a true result (eg, measures what it was intended to). **External validity**, also known as **generalizability**, is the extent to which study results can be applied outside the laboratory to real-life situations.

Lesson 2.4
Ethical Considerations

2.4.01 Informed Consent

Ethics are concerned with moral principles that guide behavior. **Ethical research** protects vulnerable populations, minimizes risks while maximizing benefits, maintains confidentiality, and respects the rights and dignity of participants. In the United States, ethics committees must approve research on human and animal subjects.

One ethical research practice involves seeking informed consent from human research participants. **Informed consent** (Figure 2.1) is the process of instructing potential research participants about the study, its risks, and their rights (eg, to withdraw, to have their information protected) so they can voluntarily decide to participate.

Figure 2.1 The process of informed consent.

Lesson 3.1
Descriptive Statistics

3.1.01 Descriptive Statistics

To analyze scientific data, researchers can use descriptive and/or inferential statistics (Lesson 3.2 covers inferential statistics). (Note: both psychologists and sociologists use descriptive and inferential statistics.)

Descriptive statistics, which describe a dataset, include **measures of central tendency** (ie, mean, median, mode) and variation (eg, range):

- **Mean** refers to the average data point. The mean represents the sum of the data points divided by the number of data points.
- **Median** refers to the middle value when data points are arranged in numerical order. The median is the midpoint at which half the data points are above and half are below. If there is an even number of data points, the two middle data points are averaged.
- **Mode** refers to the most frequently occurring data points.
- **Range** refers to the difference between the lowest and highest data points.

Lesson 3.2
Inferential Statistics

3.2.01 Correlations

Inferential statistics are used to draw conclusions that reach beyond the sample's dataset.

The Pearson correlation coefficient r can be used as an inferential statistic; the **correlation coefficient r** describes the linear relationship between two variables (Figure 3.1). The value of r ranges from −1 to 1 and describes the direction (sign) and strength of an association. The sign of the r-value indicates a positive or negative association, and the closer r is to its margins (−1 or 1), the stronger the relationship.

A positive correlation ($r > 0$) means both variables increase or decrease together. A negative correlation ($r < 0$) means that as one variable increases, the other decreases. Lastly, a lack of linear relationship between variables is called a zero correlation.

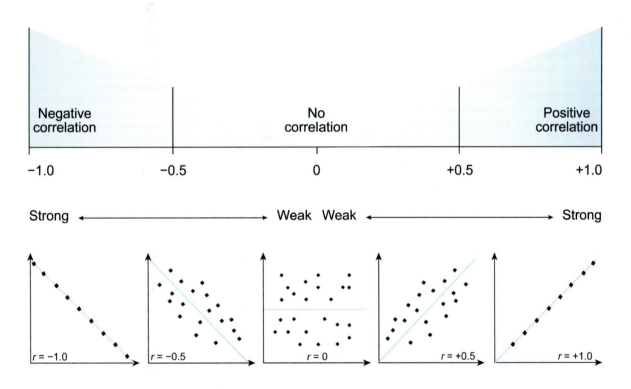

Figure 3.1 Correlation coefficient.

Because a correlation describes relationships, it does not imply causation; an r-value does not give information about the causal nature of changes in either variable.

END-OF-UNIT MCAT PRACTICE

Congratulations on completing **Unit 1: Theories and Research in Psychology**.

Now you are ready to dive into MCAT-level practice tests. At UWorld, we believe students will be fully prepared to ace the MCAT when they practice with high-quality questions in a realistic testing environment.

The UWorld Qbank will test you on questions that are fully representative of the AAMC MCAT syllabus. In addition, our MCAT-like questions are accompanied by in-depth explanations with exceptional visual aids that will help you better retain difficult MCAT concepts.

TO START YOUR MCAT PRACTICE, PROCEED AS FOLLOWS:

1) Sign up to purchase the UWorld MCAT Qbank
   IMPORTANT: You already have access if you purchased a bundled subscription.
2) Log in to your UWorld MCAT account
3) Access the MCAT Qbank section
4) Select this unit in the Qbank
5) Create a custom practice test

Unit 2 Biological Basis of Behavior

Chapter 4 The Nervous and Endocrine Systems

4.1 The Nervous System

    4.1.01   The Central Nervous System
    4.1.02   The Peripheral Nervous System

4.2 Neurons and Neural Communication

    4.2.01   Structure and Function of the Neuron
    4.2.02   Neural Communication
    4.2.03   Neurotransmitters

4.3 The Brain

    4.3.01   The Forebrain, Midbrain, and Hindbrain
    4.3.02   Lobes of the Brain
    4.3.03   Lateralization

4.4 The Spinal Cord

    4.4.01   The Spinal Cord

4.5 The Endocrine System

    4.5.01   Components of the Endocrine System
    4.5.02   Impact of the Endocrine System on Behavior

Chapter 5 Heredity and the Environment

5.1 Behavioral Genetics

    5.1.01   Adaptive Value of Traits and Behaviors
    5.1.02   Interaction of Heredity and Environmental Influences

Chapter 6 Human Physiological Development

6.1 Human Physiological Development

    6.1.01   Prenatal Development
    6.1.02   Motor Development

Chapter 7 Studying the Brain

7.1 Neuroimaging Techniques

    7.1.01   Neuroimaging Techniques

7.2 Other Methods Used in Studying the Brain

    7.2.01   Other Methods Used in Studying the Brain

Lesson 4.1

The Nervous System

4.1.01 The Central Nervous System

The **nervous system** is responsible for the regulation and integration of all body systems. Information from inside and outside the body is received and processed by the nervous system, which then coordinates purposeful reactions to this information.

The nervous system can be broadly divided into two major branches, the central and the peripheral nervous systems (Figure 4.1). The **peripheral nervous system** (PNS) comprises the neurons and glia located outside the brain and spinal cord. The **central nervous system** (CNS) consists of the brain and spinal cord and is responsible for the integration of information from the PNS. The PNS relays both sensory information to the CNS and motor commands from the CNS.

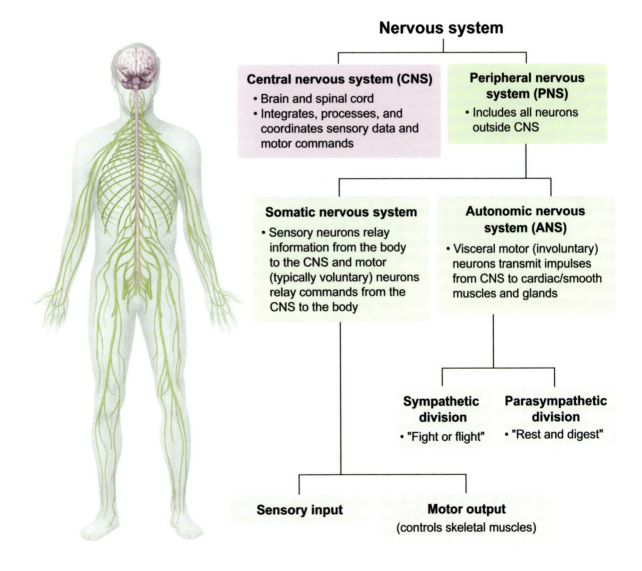

Figure 4.1 Divisions of the nervous system.

The CNS is composed of white matter and gray matter. **Gray matter** comprises unmyelinated axons, neuronal cell bodies, and dendrites. **White matter** consists primarily of myelinated axons that allow for long-distance communication between neurons.

White matter can be described as afferent (meaning toward) or efferent (meaning away from). In the context of the sensory and motor systems, **afferent** neurons send signals toward the brain, and **efferent** neurons carry messages away from the brain.

Specifically, afferent sensory neurons of the PNS relay sensory information from the body (eg, sensation from the skin) to afferent (ascending) bundles of axons of the CNS, which carry the information to the brain in the spinal cord. Efferent (descending) bundles of axons in the spinal cord carry motor commands from the brain (eg, impulses to skeletal muscles) to the body. Efferent motor neurons relay these commands to the skeletal muscles. This process is illustrated in Figure 4.2.

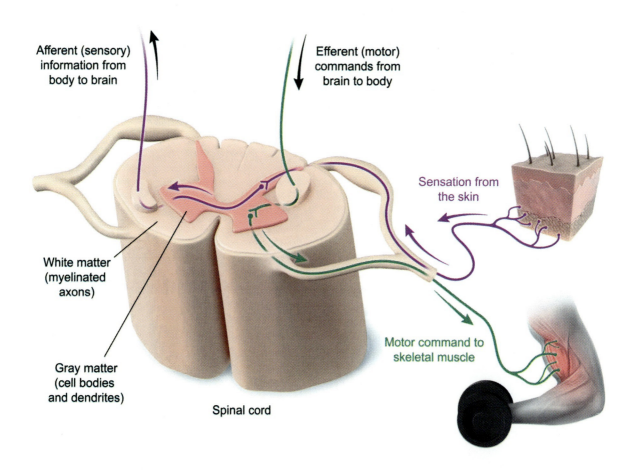

Figure 4.2 Sensory and motor neurons.

4.1.02 The Peripheral Nervous System

As introduced in Concept 4.1.01, the **peripheral nervous system** (PNS) consists of all neurons located outside the central nervous system (CNS) (ie, the brain and spinal cord). The PNS has two branches: the somatic nervous system and the autonomic nervous system (Figure 4.3).

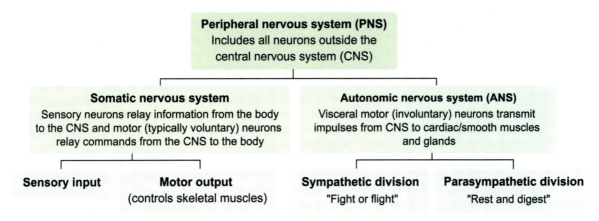

Figure 4.3 Divisions of the peripheral nervous system.

The **somatic nervous system** includes the sensory neurons that relay information from the body to the CNS and the motor neurons that relay commands from the CNS to the body.

Somatic sensory neurons transmit impulses from receptors in the skin, muscles, and joints to the spinal cord. Somatic motor neurons transmit impulses from the spinal cord to the skeletal muscles, allowing individuals to consciously perform specific movements (Figure 4.4). As a result, all movements carried out on a voluntary basis (eg, walking, talking) involve the somatic nervous system.

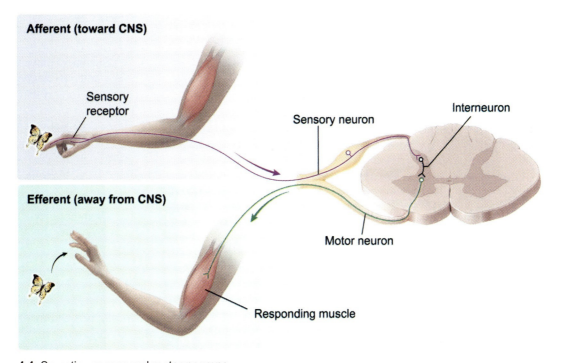

Figure 4.4 Somatic sensory and motor neurons.

In contrast, the **autonomic nervous system** (ANS) controls subconscious, automatic functions that are not subject to voluntary control. Accordingly, the activity of glands, smooth muscles, and cardiac muscles is regulated by the autonomic nervous system's visceral motor nerves. For example, the smooth muscles located within the walls of blood vessels can contract or relax to constrict or dilate the vessel.

The ANS is further divided into the **sympathetic nervous system**, which helps the body prepare for stressors, and the **parasympathetic nervous system**, which enables the body's return to homeostasis (see Figure 4.5).

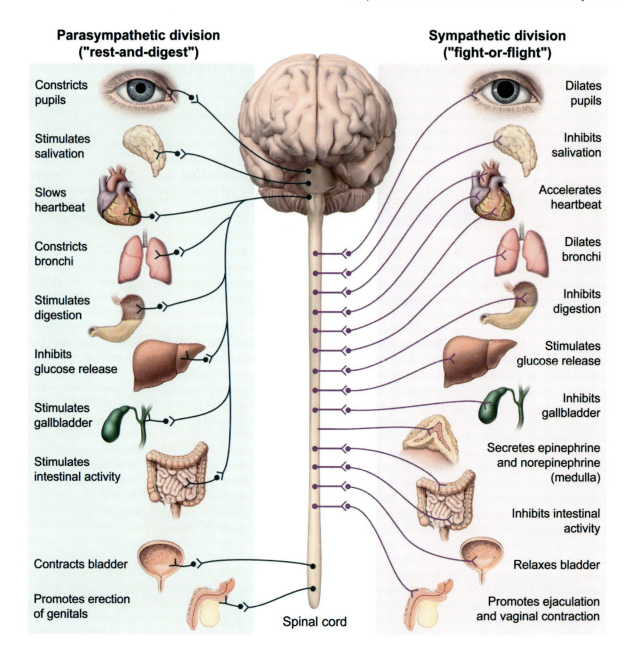

Figure 4.5 Divisions of the autonomic nervous system.

The sympathetic ("fight or flight") division mobilizes the body for action during stress. For example, oxygen delivery to the skeletal muscles is maximized (eg, increasing heart rate; dilating airways; constricting blood vessels that supply the visceral organs, glands, and skin).

The sympathetic nervous system also interacts with the endocrine system. For example, the adrenal glands are innervated by sympathetic neurons. As part of the stress response, these sympathetic neurons stimulate adrenal cells to secrete norepinephrine and epinephrine, which increase sympathetic responses (eg, blood pressure increases).

In contrast, the parasympathetic ("rest and digest") division promotes energy conservation and storage when stress is low. Parasympathetic neurons are responsible for lowering the heart rate, decreasing airflow through the lungs, increasing blood flow to the visceral organs, and promoting digestion. These functions are suppressed during times of stress.

Lesson 4.2

Neurons and Neural Communication

4.2.01 Structure and Function of the Neuron

Cells of the nervous system include both neurons and glial cells (also called glia or neuroglia). **Neurons** are responsible for sending chemical or electrical signals to other cells, and **glial cells** provide support functions to neurons and the nervous system.

A neuron's **soma** (cell body) houses the nucleus. The nucleus contains the cell's DNA, its genetic material. Everything contained within the cell is enclosed within the neuron's membrane. See Figure 4.6 for an illustration of a prototypical neuron.

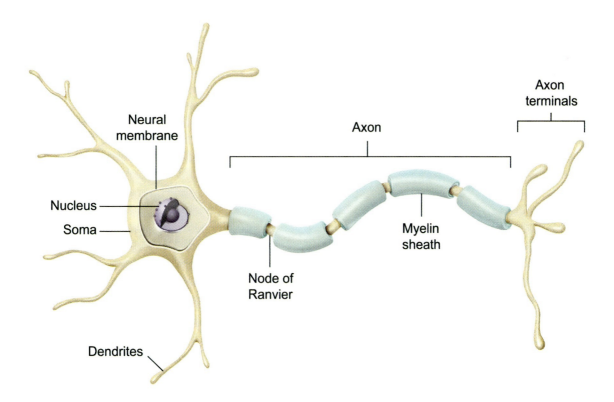

Figure 4.6 Anatomy of a prototypical neuron.

Dendrites are branches that extend from the cell body. Dendrites receive input from other neurons when neurotransmitters (chemical messengers) bind to postsynaptic receptors on the dendrites, causing the neuron to respond.

In the prototypical neuron, one **axon** then relays the neuron's output. The end of the axon at the synapse is the axon terminal (also called the terminal button). Neurons communicate when an **action potential** (AP) (electrical signal) is generated, travels down the axon, and causes the release of neurotransmitters from the axon terminal (depicted in Figure 4.7).

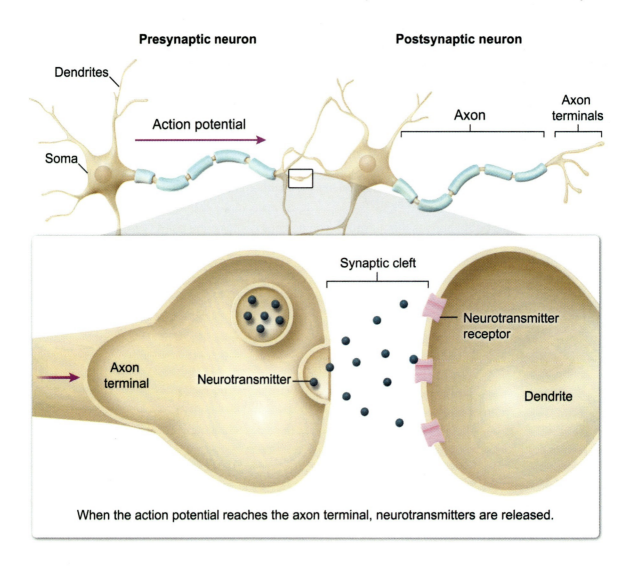

Figure 4.7 A chemical synapse.

Most neurons are myelinated, meaning that they have a layer of **myelin** (also called myelin sheath) wrapped around their axons. Myelin is a lipid, multilayered, segmented covering that surrounds and insulates axons. Produced by glia, myelin is nearly continuous along the length of an axon, interrupted only at small, regularly spaced sites called the **nodes of Ranvier**. Myelin greatly increases the speed with which the AP travels down the axon.

When the AP reaches the axon terminals (also called terminal branches), neurotransmitters are released into the **synapse** (also called the synaptic cleft), a junction formed between the axon terminal of a presynaptic neuron and the dendrites (in most cases) of a postsynaptic neuron. The neurotransmitters can then bind postsynaptic receptors, effecting change in the postsynaptic neuron.

Neural communication is aided by glia. Glial cells serve a wide range of functions; some provide structural support and chemical buffering for neurons, produce cerebrospinal fluid, or serve as immune cells in the brain. In addition, some glial cells form myelin sheaths around neurons' axons.

4.2.02 Neural Communication

Communication between neurons involves an electrical signal called an **action potential** (AP) traveling down a neuron's axon. The properties of the neuronal membrane and the ion concentrations inside and outside of the cell enable the AP.

The unequal concentration of charged ions between the inside and the outside of all living cells determines the **membrane potential** (voltage difference). The resting membrane potential (RMP) of neurons is due primarily to the high concentration of potassium ions (K^+) and the low concentration of sodium ions (Na^+) inside neurons compared to outside.

When neurotransmitters bind receptors on the dendrites of the postsynaptic neuron, it alters the membrane potential of the cell. Often, when a specific neurotransmitter binds to its receptor, ions can enter or exit the cell. See Figure 4.8 for an illustration of neurotransmitter binding.

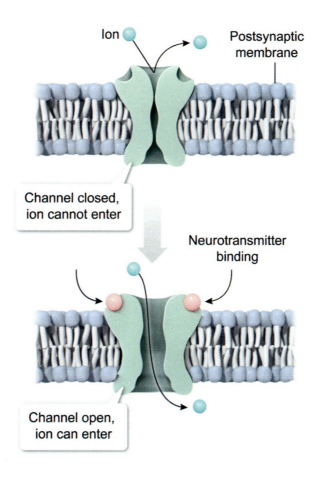

Figure 4.8 Neurotransmitter binding postsynaptic receptor.

Depending on the type of channel and the environmental conditions (eg, membrane potential) in which the channel is found, ionic movement across the membrane results in either postsynaptic depolarization (excitation) or hyperpolarization (inhibition). When neurotransmitters bind postsynaptic receptors, there is often a change in the electrical potential of the postsynaptic cell. If the change has an excitatory effect, bringing the membrane potential closer to threshold and increasing the likelihood of an AP, it is called an excitatory postsynaptic potential (EPSP). Conversely, an inhibitory postsynaptic potential (IPSP) brings the membrane potential further from threshold and decreases the likelihood of an AP.

The postsynaptic neuron receives multiple inputs; the voltage changes in electrical potential from IPSPs and EPSPs are summed. If the summed EPSPs and IPSPs from presynaptic neurons cause depolarization of the postsynaptic neuron that exceeds a certain threshold, an AP is fired (see Figure 4.9).

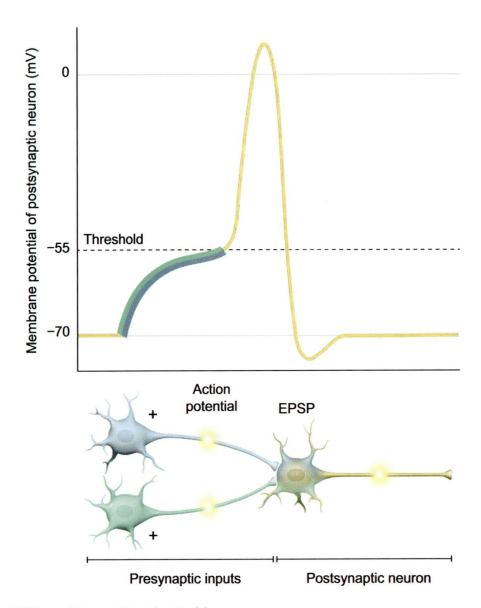

EPSP = excitatory postsynaptic potential.

Figure 4.9 Summation of postsynaptic potentials.

The AP is a change in the membrane potential of a neuron and results in the transmission of an electrical signal down its axon. During an AP, the membrane potential of the neuron changes due to the opening and closing of Na^+ and K^+ ion channels, as follows:

- At rest, postsynaptic Na^+ channels are closed, but if a stimulus causes them to open, Na^+ ions enter the cell, causing the membrane to depolarize (ie, the membrane potential becomes more positive).
- If this depolarization is enough to reach a certain threshold, an AP is initiated in the neuron. If the threshold is reached, additional Na^+ channels open, causing further, rapid depolarization of the membrane.
- Repolarization occurs when Na^+ channels inactivate and K^+ channels open, allowing K^+ to rush out of the cell, and the membrane potential becomes negative again.
- The RMP is fully restored when K^+ and Na^+ channels are closed.

These steps are shown in Figure 4.10.

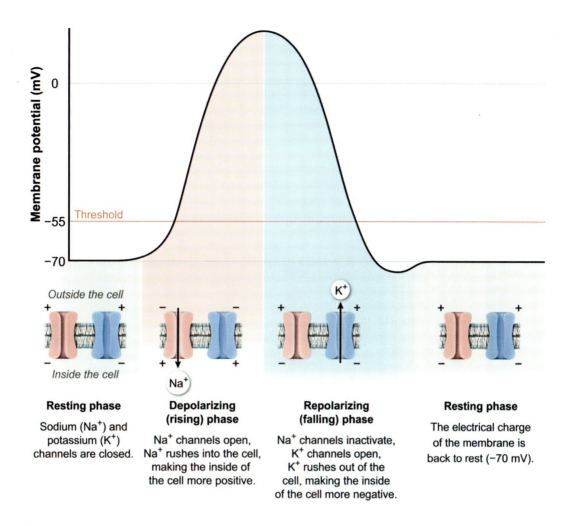

Figure 4.10 Phases of the action potential.

Another AP cannot occur during repolarization, the phase during which K⁺ channels open and Na⁺ channels inactivate. The **absolute refractory period** refers to this time, during which no new APs can be generated, regardless of the strength of the stimulus received.

APs travel down an axon toward the axon terminal as successive regions are depolarized. The voltage-gated Na⁺ and K⁺ channels open and close in a sequential manner to propagate the electrical signal down the axon (Figure 4.11).

Chapter 4: The Nervous and Endocrine Systems

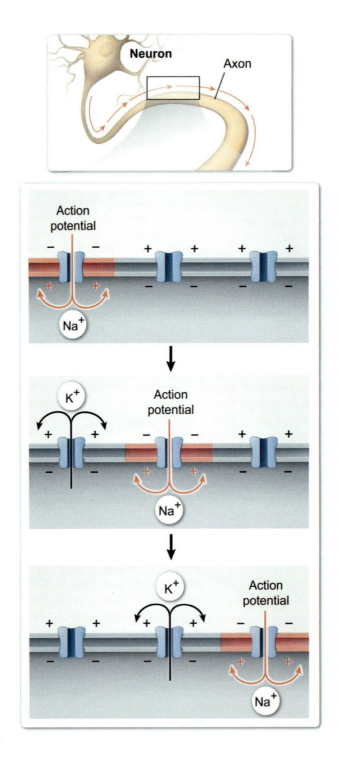

Figure 4.11 Action potential propagation.

The rate at which an AP travels down the axon is influenced by axon size, as well as by myelination; larger-diameter axons and axons that are myelinated transmit neural impulses faster than smaller and unmyelinated axons, respectively.

At chemical synapses, when an AP reaches the axon terminal of the presynaptic neuron, neurotransmitters are released from the synaptic vesicles (where they are stored) into the synaptic cleft. The neurotransmitters travel across the synaptic cleft, and some bind to receptors on the postsynaptic cell membrane.

In most cases when a neurotransmitter binds to its receptor on the postsynaptic neuron (Figure 4.12), ions can move through the open channel into or out of the postsynaptic neuron, which alters the neuron's RMP. The type of receptor and the postsynaptic neuron's membrane potential determine whether ionic movement across the membrane results in either postsynaptic excitation (ie, depolarization) or inhibition.

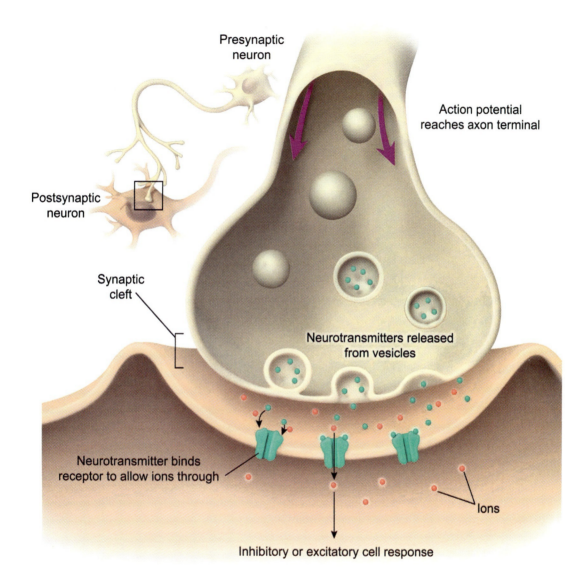

Figure 4.12 Neurotransmitter release at the synapse.

Depending on the net effect on the postsynaptic cell, synapses can be classified as excitatory (promoting AP initiation) or inhibitory (inhibiting AP initiation). At excitatory synapses, the membrane potential of the postsynaptic neuron becomes more positive (ie, depolarizes), and if it exceeds threshold, an AP is initiated.

In contrast, at inhibitory synapses, neurotransmitters affect the postsynaptic neuron by causing the cell's membrane potential to become more negative, inhibiting AP initiation. An excitatory and an inhibitory synapse are contrasted in Figure 4.13.

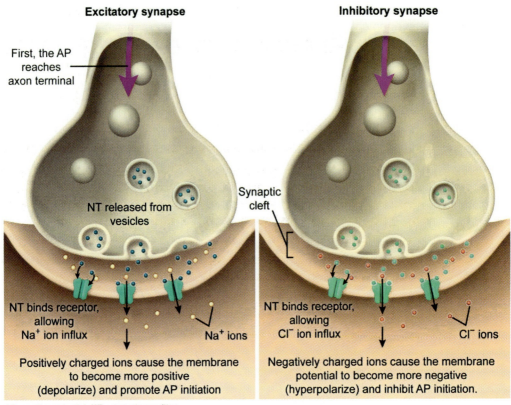

Figure 4.13 Example of an excitatory versus an inhibitory synapse.

Neurotransmitters themselves can also be classified as excitatory (ie, they make an AP more likely to occur in the postsynaptic neuron), or inhibitory (ie, they make an AP less likely to occur in the postsynaptic neuron).

4.2.03 Neurotransmitters

The amino acid neurotransmitters are gamma-aminobutyric acid (GABA), glutamate, and glycine. **GABA** is the principal inhibitory neurotransmitter of the CNS, and **glycine** is also a primarily inhibitory neurotransmitter found in the CNS. **Glutamate** is the primary excitatory neurotransmitter of the CNS. Figure 4.14 shows examples of synapses using glutamate and GABA.

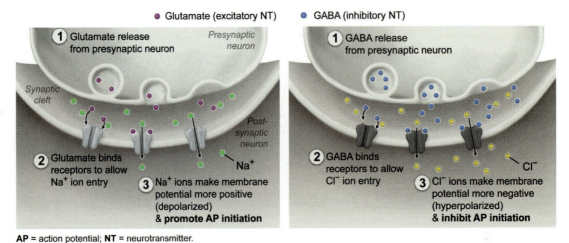

Figure 4.14 Glutamate and GABA at the synapse.

Acetylcholine is the neurotransmitter released by somatic motor neurons at the neuromuscular junction. In addition, acetylcholine plays an important role in the sympathetic and parasympathetic nervous systems. In the brain, acetylcholine is involved in learning and memory, arousal, and sleep-wake cycles.

The catecholamines are a category of neurotransmitter that includes dopamine, epinephrine (also called adrenaline), and norepinephrine (also called noradrenaline). **Dopamine** is involved in movement, reward, motivation, and cognition. The dopaminergic system is also implicated in schizophrenia and Parkinson's disease and is discussed in Concept 30.1.01 and Lesson 30.2, respectively. **Epinephrine** and **norepinephrine** are involved in autonomic nervous system responses and contribute to wakefulness, alertness, attention, memory, and mood.

The monoaminergic neurotransmitters include the catecholamines and serotonin. **Serotonin** is a neurotransmitter involved in sleep/wake regulation, mood, and appetite, among others.

Endorphins are peptides that act as neurotransmitters. Endorphins are opioids produced by the body that, similar to morphine, modulate pain, as well as contribute to elevated mood following exercise.

The neurotransmitter classifications are shown in Table 4.1.

Table 4.1 Primary neurotransmitters.

	Neurotransmitter	**Functions**
Amino acids	Glutamate (Glu)	• Primary excitatory neurotransmitter of the central nervous system • Involved in learning and memory
	Gamma-aminobutyric acid (GABA)	• Primary inhibitory neurotransmitter of the brain
	Glycine (Gly)	• Primary inhibitory neurotransmitter of the spinal cord
Amines	Dopamine (DA)	• Involved in cognition, attention, movement, reward
	Serotonin (5-HT)	• Involved in sleep, appetite, mood
	Epinephrine	• Involved in sympathetic signaling in the autonomic nervous system
	Norepinephrine (NE)	• Involved in sympathetic signaling in the autonomic nervous system
	Acetylcholine (ACh)	• Involved in parasympathetic signaling in the autonomic nervous system • Released by motor neurons at NMJs of the somatic nervous system to excite skeletal muscle
Peptides	Endorphins	• Opioids produced by the body that modulate pain, as well as contribute to elevated mood following exercise

NMJ *= neuromuscular junction.*

Excess neurotransmitters can diffuse out of the cleft, get broken down, or get reabsorbed back into the presynaptic neuron through reuptake. Some drugs affect these processes, altering neural communication. **Agonists** mimic or enhance the effects of a neurotransmitter (eg, by inhibiting reuptake). **Antagonists** block or reduce the effects of a neurotransmitter (eg, by blocking postsynaptic receptors). Figure 4.15 depicts agonists and antagonists.

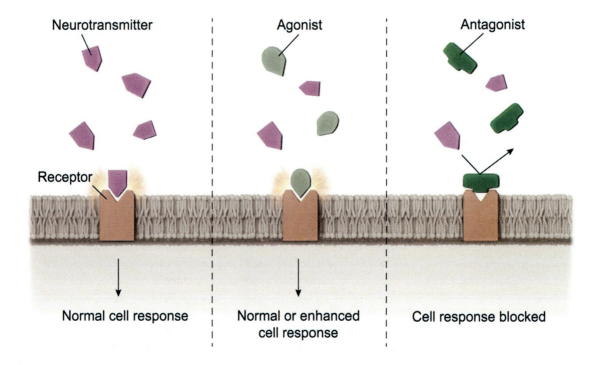

Figure 4.15 Example of one type of agonist and antagonist.

Lesson 4.3
The Brain

4.3.01 The Forebrain, Midbrain, and Hindbrain

The brain can be described as comprising three major regions: the forebrain, the midbrain, and the hindbrain.

The **forebrain** (prosencephalon) contains the cerebrum, the brain's two hemispheres. Major forebrain structures include the olfactory bulbs (discussed in more detail in Lesson 12.3), the basal ganglia (a collection of nuclei involved in initiating voluntary movements), and the pineal gland. The pineal gland releases melatonin, a hormone that causes sleepiness, when it is dark (see Lesson 13.2).

The forebrain also includes several brain structures that contribute to emotion, memory, and motivation. These structures are sometimes collectively referred to as the limbic system (Figure 4.16). Limbic structures include:

- The **amygdala** is involved in aggression and emotions such as fear. Researchers have shown that electrical stimulation of the amygdala leads to displays of fear and aggression, whereas damage to the amygdala can result in a lack of fear.
- The **hypothalamus** releases hormones and controls the pituitary gland's hormone release; it coordinates many bodily processes such as hunger, growth, and the fight-or-flight stress response, as is discussed in Lesson 4.5.
- The **hippocampus** is involved in learning and memory, such as the formation of explicit/declarative memories (ie, memory for facts and events that can be intentionally recalled) (Lesson 19.3 covers the biological underpinnings of memory in further detail). Damage to the hippocampus (eg, brain trauma caused by a head injury) can result in amnesia: severe memory loss.
- The **thalamus** contributes to sensation and perception; it is responsible for processing and relaying sensory information and directly receives information from all the senses except olfaction.

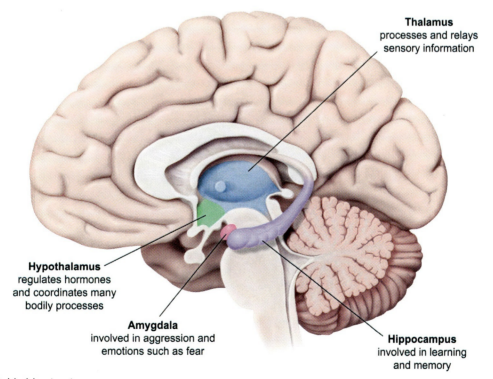

Figure 4.16 Limbic structures.

The **midbrain** (mesencephalon) is an area of the brainstem that connects the brain and the spinal cord. The superior and inferior colliculi, structures located in the midbrain, serve important sensory functions; the superior colliculus processes visual information, and the inferior colliculus processes auditory information.

In addition, the midbrain contains two areas with large numbers of dopaminergic neurons: the substantia nigra (SN) and the ventral tegmental area (VTA). The SN projects axons to the basal ganglia and plays an important role in voluntary movements. Parkinson's disease, a progressive neurodegenerative disease, is marked by the death of dopaminergic neurons in the SN and results in impaired movement (see Concept 29.2.01). The VTA projects to different parts of the forebrain (eg, prefrontal cortex, nucleus accumbens) and plays an important role in reward.

Finally, the **hindbrain** (rhombencephalon) is made up of the cerebellum and the lower part of the brainstem, including the medulla, pons, and reticular formation (Figure 4.17).

The brainstem region closest to the spinal cord is the **medulla**, which controls critical functions such as breathing and heart rate. The **pons** lies above the medulla and regulates sleeping, waking, and dreaming. The **reticular formation** is a series of neurons that spans the entire brainstem and contributes to consciousness and wakefulness.

Located behind the brainstem, the **cerebellum** is responsible for coordinating voluntary movements, posture, and balance. Specifically, the cerebellum integrates visual, vestibular, and kinesthetic information to maintain balance and posture, coordinate complex movements, and execute precise fine motor movements. The cerebellum is also critical for motor learning, which occurs when an organism repeatedly practices a motor task (eg, swimming).

Pons
- Brainstem structure
- Regulates sleeping, waking, and dreaming

Medulla
- Brainstem structure
- Controls critical functions such as breathing and heart rate

Reticular formation
- A series of neurons that spans the entire brainstem
- Contributes to consciousness and wakefulness

Cerebellum
- Involved in motor learning, movement coordination, and balance

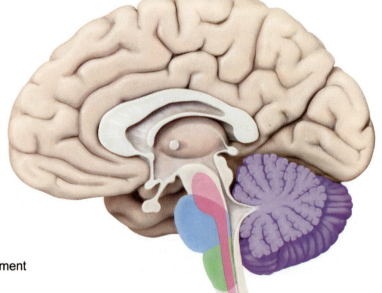

Figure 4.17 Hindbrain structures.

4.3.02 Lobes of the Brain

The **cerebral cortex** is the outermost portion of the brain. The cortex can be divided into the frontal, temporal, parietal, and occipital lobes (Figure 4.18).

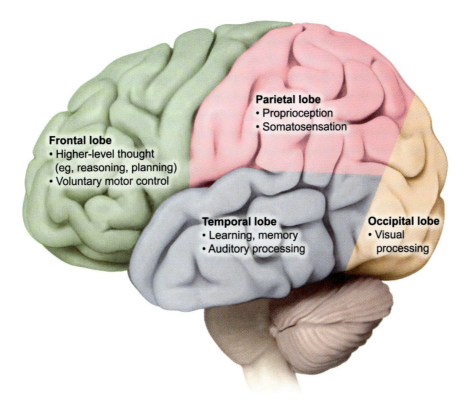

Figure 4.18 Lobes of the brain.

The frontal lobe is responsible for higher-order processes (eg, planning, decision-making, personality, judgment). The **prefrontal cortex** is an area of the frontal lobe that contributes to decision-making, personality, and memory.

Voluntary muscle movements also involve the frontal lobe. The frontal lobe's **motor cortex** relays motor commands from the motor cortex to the skeletal muscles. The regions of the body requiring more motor control (eg, the hands) occupy a greater area of the motor cortex.

The temporal lobe contains the **auditory cortex**, an area primarily responsible for processing auditory stimuli based on input from the ears (see Lesson 11.2). The temporal lobe also contains structures involved in learning and memory (eg, the hippocampus), as well as areas involved in language (eg, Wernicke's area).

The parietal lobe is responsible for proprioception (awareness of one's body position in space) and somatosensation (eg, perception of touch, pain, temperature). Specifically, somatosensory information is processed in the parietal lobe's somatosensory cortex. The more sensitive regions of the body (eg, fingers, tongue) occupy a greater area of the somatosensory cortex. See Chapter 12 for more information on the role of the parietal lobe in somatosensation and proprioception.

The occipital lobe is located in the back of the cerebral cortex and is responsible primarily for visual processing. The occipital lobe's **visual cortex** receives and processes input from the eyes. Visual processing is discussed in more detail in Lesson 10.2.

4.3.03 Lateralization

The brain's right and left hemispheres are each specialized for certain processes; this specialization is known as **hemispheric lateralization**. Each hemisphere is responsible for contralateral control of the body; the left hemisphere controls touch and movement on the right side of the body, and vice versa.

In the majority of people, the right hemisphere is crucial for visuospatial processing. In contrast, the left hemisphere in most people is specialized for language functions, including writing, speech production (Broca's area), and language comprehension (Wernicke's area). These language centers are shown in Figure 4.19.

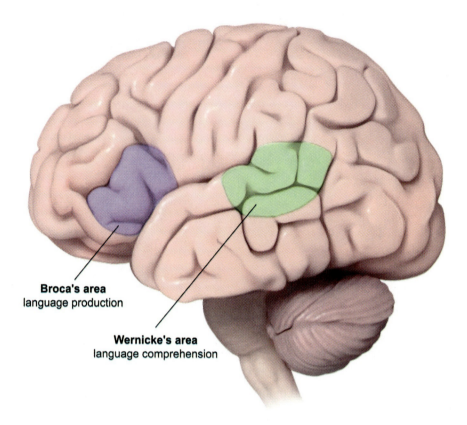

Figure 4.19 Language centers in the brain.

The **corpus callosum** is a bundle of myelinated axonal projections connecting the right and left hemispheres of the brain, allowing the two hemispheres to communicate. Severing the corpus callosum is sometimes used to treat severe epileptic seizures. Individuals with a severed corpus callosum ("split-brain") experience disrupted communication between the two hemispheres.

In famous experiments, Roger Sperry and Michael Gazzaniga demonstrated that information presented to a split-brain patient's left visual field is processed in the right hemisphere; without interhemispheric communication, the patient is unable to express what is seen verbally but would be able to draw it with the left hand (Figure 4.20).

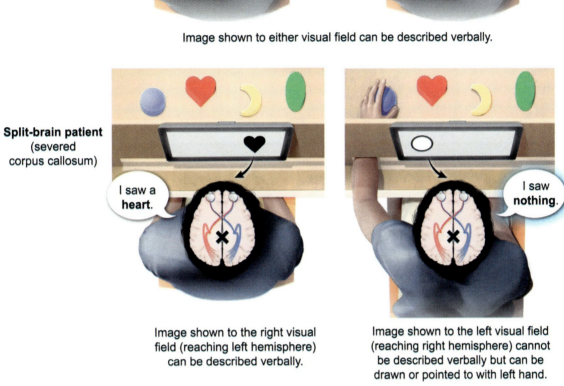

Figure 4.20 Split-brain research.

Lesson 4.4

The Spinal Cord

4.4.01 The Spinal Cord

The **spinal cord** is a structure in the central nervous system that facilitates the brain's communication with the peripheral nervous system (PNS). While the spinal cord relays messages between the PNS and the brain, information processing also occurs in the spinal cord.

The spinal cord contains tracts of white matter (myelinated axons) and gray matter (cell bodies and dendrites) as discussed in Concept 4.1.01. In the context of the sensory and motor systems, the spinal cord's afferent (ascending) tracts send sensory signals toward the brain, and efferent (descending) tracts carry motor commands away from the brain. The organization of the spinal cord is illustrated in Figure 4.21.

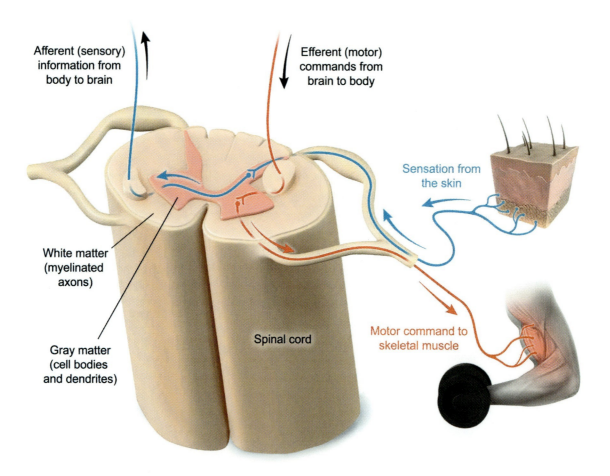

Figure 4.21 Information relayed through the spinal cord.

In addition to transmitting sensory and motor information to the brain, some spinal cord neurons are responsible for processing information. An example of this is a **spinal reflex**.

The neural process of receiving and acting on sensory information in this reflex arc involves the following steps:

1. Somatosensory receptors in the skin are stimulated by something painful (eg, a prick from a sharp needle).
2. Afferent sensory neurons relay this information to spinal interneurons (neurons in the spinal cord that integrate sensory and central nervous system inputs).
3. Spinal interneurons process the sensory information before it travels to the brain and directly stimulate efferent motor neurons.
4. Efferent motor neurons relay motor commands to the skeletal muscles, causing the muscles to respond (eg, pull the hand away from the needle).

The steps of this reflex are shown in Figure 4.22.

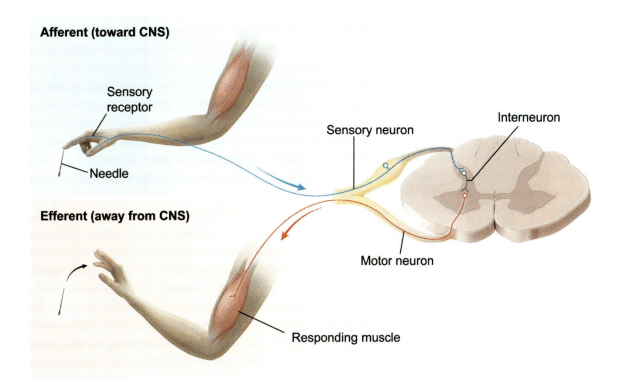

Figure 4.22 A spinal reflex.

Lesson 4.5

The Endocrine System

4.5.01 Components of the Endocrine System

In addition to the nervous system, the body also communicates through the **endocrine system**. In contrast to neural communication wherein cells communicate by forming synapses with nearby cells, endocrine communication is a slower form of long-distance cell-to-cell communication.

The endocrine system regulates physiological activity through the secretion of **hormones**, chemical messengers that travel throughout the body via the bloodstream. Hormones then bind receptors in target tissues to elicit specific responses (eg, altering cellular function).

Hormones affect the function of diverse and distant tissues because hormone receptors can be found on various cell types throughout the body. Any cell with the correct receptor is capable of responding to the hormone.

Hormones are regulated by a brain structure called the hypothalamus. The **hypothalamus** serves important functions for both the nervous and endocrine systems by processing inputs from the cortex and sensing the plasma concentration of numerous hormones. In response to these inputs, the hypothalamus controls body-wide endocrine function by releasing hormones and regulating hormone release from the **pituitary gland**, an endocrine organ located below the hypothalamus (see Figure 4.23).

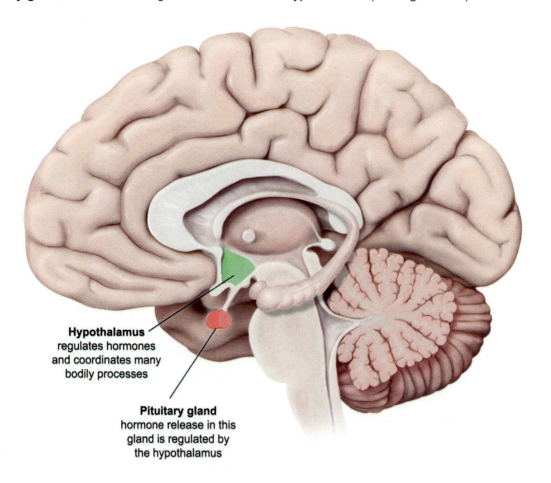

Figure 4.23 The hypothalamus and pituitary gland.

Together, the hypothalamus and the pituitary gland regulate hormone release from the rest of the endocrine system's glands. An overview of the endocrine system is shown in Figure 4.24.

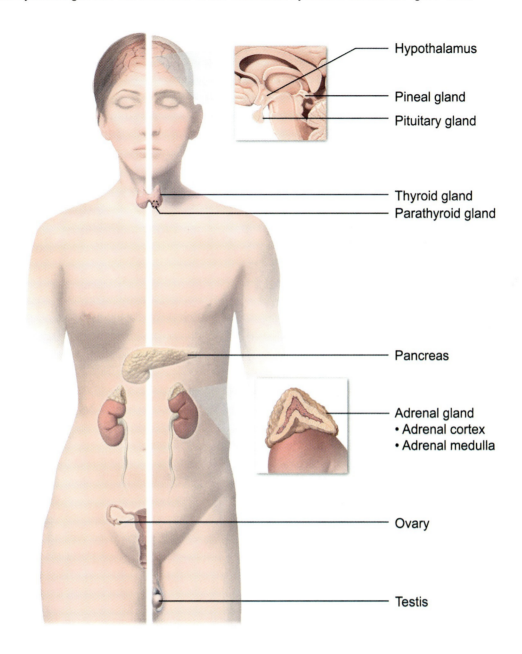

Figure 4.24 The endocrine system.

4.5.02 Impact of the Endocrine System on Behavior

Hypothalamic and pituitary hormones coordinate many bodily processes, such as growth, blood pressure, core body temperature, appetite, sleep, and the stress response.

The hypothalamus has several nuclei (collections of neuronal cell bodies) with specialized functions. One of these nuclei is the **suprachiasmatic nucleus**, which is the area that regulates circadian rhythms, cycles in physiological activity or behavior that occur over 24-hour intervals (eg, the sleep/wake cycle). Another hypothalamic nucleus, the ventromedial nucleus, regulates hunger and satiety.

The pituitary gland can be divided into two separate lobes, the **anterior pituitary** and the **posterior pituitary**. The two lobes release distinct hormones involved in regulating different bodily processes. One example is oxytocin, a hormone released by the posterior pituitary gland that is involved in pair bonding, reproductive behavior, labor, and lactation.

The hypothalamus and pituitary also impact behavior by affecting endocrine glands throughout the body. For example, during stress, the hypothalamus releases a hormone that causes the pituitary gland to release adrenocorticotropic hormone (ACTH), a stress hormone. ACTH travels in the bloodstream to the adrenal glands (endocrine organs on top of the kidneys), where it causes the release of additional stress hormones (eg, cortisol) that help activate the body to deal with the stressor (Figure 4.25).

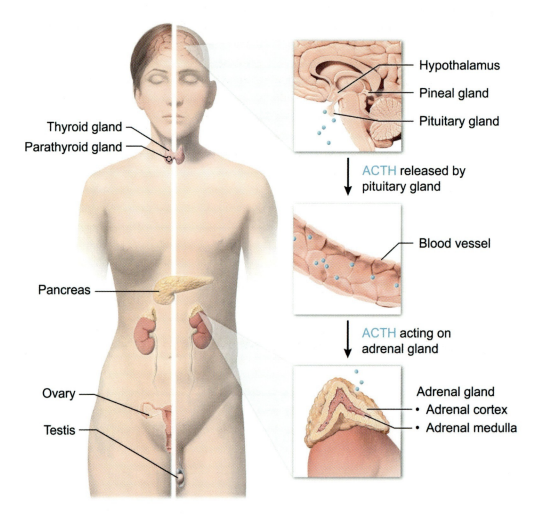

Figure 4.25 Endocrine hormones acting on the adrenal glands.

Lesson 5.1
Behavioral Genetics

5.1.01 Adaptive Value of Traits and Behaviors

Adaptive value refers to the extent to which a trait or behavior helps an organism survive and reproduce. Traits and behaviors that are innate (ie, unlearned) result from genetic influences; babies are born with many **reflexes** (eg, feeding reflexes like suckling and rooting) that are preprogrammed behaviors which help them survive.

An additional example is the research finding that infants prefer human faces and human speech, suggesting that this innate preference has evolved. This is an adaptive behavior because recognition of these stimuli confers a survival advantage: faces and speech convey important social information (eg, emotion) and contribute to the pair-bonding process between infant and caregiver.

Conversely, behaviors that are learned (ie, nongenetic) result from observation and experience. They can change over time with practice or environmental demands. For example, feeding oneself with utensils (eg, spoon, chopsticks) is a learned behavior. Most human behaviors involve both genetic and environmental contributions, falling along the continuum from innate to learned (Figure 5.1).

Figure 5.1 Continuum of human behavior.

5.1.02 Interaction of Heredity and Environmental Influences

Physical and social development are influenced by both heredity and the environment. **Heredity**, sometimes referred to as nature, describes genetic influences on development (eg, genes coding for eye color). In contrast, the **environment**, sometimes referred to as nurture, describes all nongenetic influences (eg, parenting styles).

Many traits are a result of the interaction of nature and nurture. For example, an individual's height is determined by both the genetic information passed down from their biological parents as well as the influence of their environment, such as the nutrition they received as a child.

Because heredity and the environment both impact traits and behavior in significant ways, researchers use twin studies and adoption studies to estimate the relative contribution of genetic versus environmental factors. Although they are rare, **twin adoption studies** can help clarify the role of heredity and the environment for complex human traits.

Identical (monozygotic) twins raised together share the same genes and an extremely similar environment (eg, same household, same schools, similar experiences), so it is not possible to determine if similar traits are the result of genetics, environment, or a combination of the two.

However, if identical twins are each adopted and raised apart, traits that they share are most likely determined by genetics, whereas traits that are more similar to those of their adoptive families are most likely determined by environmental influences (see Figure 5.2).

Identical twins raised together

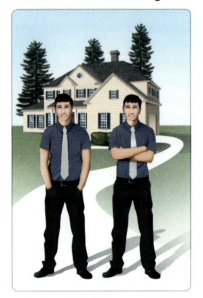

100% shared genes
Shared environment

Identical twins raised apart

100% shared genes
Different environment

Figure 5.2 Identical twins raised together versus apart.

Lesson 6.1

Human Physiological Development

6.1.01 Prenatal Development

The prenatal period of development occurs during gestation, the time between conception and birth. Prenatal development is influenced by both nature and nurture. Nature refers to genetic influences on development (eg, genes coding for eye color), whereas nurture refers to nongenetic influences in the environment (eg, exposure to nicotine). For example, one's height is determined by both the genetic information passed down from their parents and the influence of their environment, such as the nutrition they received during gestation.

A teratogen is any factor (eg, a drug or virus) that negatively impacts prenatal development. For example, consumption of alcohol, a teratogen, during pregnancy can lead to fetal alcohol syndrome. **Fetal alcohol syndrome** causes permanent negative effects to the fetus such as inhibited growth, facial deformities, and damage to the brain, resulting in lower intelligence.

6.1.02 Motor Development

Humans are born with simple **reflexes**: automatic responses to sensory stimuli that aid in survival. Some of these reflexes include:

- The Moro reflex (also called the startle reflex), which causes an infant to throw back their head, extend their arms and legs, cry, and clench their arms and legs to their body when startled
- Rooting, which causes an infant to turn their head and open their mouth when stroked on the cheek
- The patellar reflex (also called the knee-jerk reflex), which causes the leg to extend when the patellar tendon (below the kneecap) is tapped
- The Babinski reflex, which causes an infant to bend their big toe back while their other toes fan outward when the bottom of their foot is stroked

In contrast, motor skills are voluntary movements of the body. Over time, complex motor skills develop in a common, predictable order (eg, rolling over before sitting, sitting before standing). **Gross motor skills** involve large muscle movements (eg, waving an arm). **Fine motor skills** involve smaller muscle movements that allow a person to perform more precise actions (eg, pinching an object between the thumb and index finger).

Gross motor skills develop before fine motor skills (eg, children can wave their arms before they can hold a crayon). By elementary school age (~6 years), children can easily grasp and manipulate small items (eg, buttons).

Lesson 7.1

Neuroimaging Techniques

7.1.01 Neuroimaging Techniques

Neuroimaging techniques detect brain structure and/or function (see Table 7.1).

Neuroimaging that detects *brain structure* reveals the size of brain areas and surrounding structures (eg, the fluid-filled ventricles of the brain), as well as any abnormalities in the tissue (eg, tumors). Techniques that are specialized for visualizing brain structures include:

- Computerized tomography (CT) or computerized axial tomography (CAT) generates images of the brain by combining x-ray images taken from different angles.
- Magnetic resonance imaging (MRI) uses powerful magnets to create detailed images of the brain. MRI provides a closer and more precise image than a CT scan.

Other neuroimaging techniques reveal *brain activity* and *functioning*. Techniques that are useful in assessing brain function include:

- Positron emission tomography (PET) provides information about physiological activity in the brain by monitoring glucose metabolism. Active neurons consume glucose (sugar) for energy, and more active brain regions use more glucose. PET measures positively charged particles (positrons) that are emitted during the metabolism of a radioactive glucose tracer injected prior to the scan.
- Functional magnetic resonance imaging (fMRI) is used to visualize brain activity by measuring changes in blood oxygen levels in the brain. Active brain areas require more oxygen and an increased blood supply. fMRI uses a powerful magnet to detect which areas of the brain require increased blood flow, indicating increased brain activity.
- Electroencephalography (EEG) measures electrical brain activity through the use of electrodes attached to the scalp. Because action potentials involve a sequence of shifts in the electrical charge of the neural membrane, neural communication involves electrical activity. The readout from an EEG depicts waves of brain activity (eg, showing stages of sleep).

Table 7.1 Selected neuroimaging methods.

Neuroimaging method	Protocol	Provides information about
Computerized tomography (CT)	Computer combines multiple X-rays taken at different angles	Structure of internal organs and tissues at a single point in time
Magnetic resonance imaging (MRI)	Powerful magnets create detailed images	Structure of internal organs and tissues at a single point in time
Positron emission tomography (PET)	Scanner detects radioactive glucose tracer	Changes in glucose metabolism in the brain over time
Functional magnetic resonance imaging (fMRI)	Scanner detects the differential properties of blood that is high versus low in oxygen	Changes in blood oxygenation in the brain over time
Electroencephalography (EEG)	Electrodes placed on the scalp	Electrical fluctuations in the brain over time

Lesson 7.2
Other Methods Used in Studying the Brain

7.2.01 Other Methods Used in Studying the Brain

In addition to neuroimaging (see Lesson 7.1), scientists use a variety of other techniques in their study of the brain.

Electrical stimulation of the brain (ESB) stimulates precise brain areas with an electric current. For example, stimulation of the region of the motor cortex in the left hemisphere that controls the hand would cause movement of the right hand. ESB has many applications; for instance, it can be used to determine the function of specific brain areas.

Another technique, **lesioning**, involves destroying specific areas of the brain. Lesioning is a research tool used in studies with animals. For example, scientists could use lesioning to learn about the function of a brain area by determining what processes are disrupted as a result of damage to that area. Lesioning can also be used to treat movement disorders such as dystonia (ie, involuntary movement).

END-OF-UNIT MCAT PRACTICE

Congratulations on completing **Unit 2: Biological Basis of Behavior**.

Now you are ready to dive into MCAT-level practice tests. At UWorld, we believe students will be fully prepared to ace the MCAT when they practice with high-quality questions in a realistic testing environment.

The UWorld Qbank will test you on questions that are fully representative of the AAMC MCAT syllabus. In addition, our MCAT-like questions are accompanied by in-depth explanations with exceptional visual aids that will help you better retain difficult MCAT concepts.

TO START YOUR MCAT PRACTICE, PROCEED AS FOLLOWS:

1) Sign up to purchase the UWorld MCAT Qbank
   IMPORTANT: You already have access if you purchased a bundled subscription.
2) Log in to your UWorld MCAT account
3) Access the MCAT Qbank section
4) Select this unit in the Qbank
5) Create a custom practice test

Unit 3 Sensation, Perception, and Consciousness

Chapter 8 Sensation

8.1 Principles of Sensation

- 8.1.01 Sensory Adaptation
- 8.1.02 Psychophysics

8.2 Sensory Receptors

- 8.2.01 Types of Sensory Receptors
- 8.2.02 Sensory Pathways

Chapter 9 Perception

9.1 Principles of Perception

- 9.1.01 Top-Down and Bottom-Up Processing
- 9.1.02 Perceptual Organization
- 9.1.03 Gestalt Principles

Chapter 10 Vision

10.1 Eye Structure and Function

- 10.1.01 Eye Structure and Function

10.2 Visual Processing

- 10.2.01 Visual Processing

Chapter 11 Hearing

11.1 Ear Structure and Function

- 11.1.01 Ear Structure and Function

11.2 Auditory Processing

- 11.2.01 Auditory Processing

Chapter 12 Other Senses

12.1 Somatosensation

- 12.1.01 The Biological Underpinnings of Somatosensation
- 12.1.02 The Process of Somatosensation

12.2 Taste

- 12.2.01 The Biological Underpinnings of Taste
- 12.2.02 The Process of Taste

12.3 Smell

- 12.3.01 The Biological Underpinnings of Smell
- 12.3.02 The Process of Smell

12.4 The Kinesthetic Sense

    12.4.01   The Biological Underpinnings of the Kinesthetic Sense
    12.4.02   The Kinesthetic Process

12.5 The Vestibular Sense

    12.5.01   The Biological Underpinnings of the Vestibular Sense
    12.5.02   The Vestibular Process

Chapter 13 Consciousness and Sleep

13.1 States of Consciousness

    13.1.01   Alertness
    13.1.02   Sleep

13.2 The Stages of Sleep

    13.2.01   Sleep Cycles
    13.2.02   Circadian Rhythms
    13.2.03   Dreaming
    13.2.04   Sleep-Wake Disorders

Chapter 14 Consciousness-Altering Substances

14.1 Consciousness-Altering Substances

    14.1.01   Consciousness-Altering Substances

14.2 Problematic Substance Use

    14.2.01   Problematic Substance Use

Lesson 8.1

Principles of Sensation

8.1.01 Sensory Adaptation

Sensation arises through the process of transduction. **Transduction** occurs when sensory receptors (eg, olfactory receptors) convert environmental stimuli (eg, a smell) into neural signals. The nervous system then relays that information to different areas of the brain for processing.

Sensory adaptation occurs when the constant presence of a sensory stimulus causes sensory receptor cells to send fewer messages to the brain about that stimulus. In other words, when a stimulus does not change, the firing rate of the responding receptor declines. Sensory adaptation explains why, for example, an individual would initially detect a strong garbage odor but, after a few hours, be less able to detect the smell (see Figure 8.1).

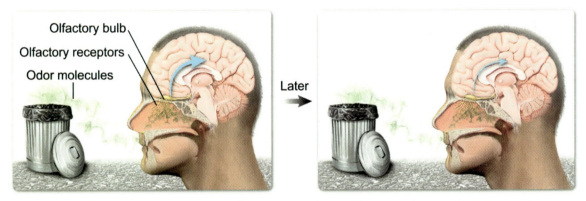

For example, a strong garbage smell (ie, sensory stimulus) becomes less noticeable over time.

Figure 8.1 Sensory adaptation example.

8.1.02 Psychophysics

In the 1800s, Ernst Weber and Gustav Fechner studied how sensation could be altered by varying the intensity of a stimulus.

Ernst Weber studied the **difference threshold**, also called the **just-noticeable difference** (JND), the point at which an individual can detect a difference between two stimuli (eg, the heaviness of two weights) 50% of the time (Figure 8.2).

Figure 8.2 Example of the difference threshold.

As the intensity of the stimulus increases, the amount of change needed to detect a difference in that stimulus also increases. This relationship is quantified by **Weber's law**, which states that the proportion of the size of the JND to the original stimulus intensity is a constant:

$$\frac{\Delta I}{I} = k$$

where ΔI is the size of the JND, I is the original stimulus intensity, and k is a proportionality constant. Once the JND has been determined for a particular stimulus intensity, it can be predicted for new stimulus intensities.

For example, consider that a weight of 30 lb must be increased by 6 lb for an individual to notice a difference in heaviness between the two weights half the time. In this example,

$$\frac{6 \text{ lb}}{30 \text{ lb}} = 0.2$$

If a 40-lb weight is contrasted with progressively heavier weights, this formula will determine the lowest weight at which the individual can detect the increase. The new stimulus intensity for the weight is 40 lb, and the new JND can be calculated by

$$\frac{\Delta I}{40 \text{ lb}} = 0.2$$
$$\Delta I = 8 \text{ lb}$$

Therefore, the 40-lb weight must be increased to 40 lb + 8 lb = 48 lb for a difference in heaviness to be detected.

Another scientist, **Gustav Fechner**, studied the **absolute threshold**, the point at which an individual can detect a new sensation (eg, see a light in the distance) 50% of the time. Figure 8.3 depicts common absolute thresholds.

For example, common absolute thresholds include being able to:

Hear a watch ticking 20 feet away

See a candle flame on a clear night 30 miles away

Taste 1 teaspoon of sugar dissolved in 2 gallons of water

Feel a bee's wing falling from 1 cm onto cheek

Smell 1 drop of perfume in a 3-room house

Figure 8.3 Examples of the absolute threshold.

The ability to correctly detect a stimulus is also impacted by the environment. **Signal detection theory** explores how judgments or decisions are made amid "noise" (external or internal distractions).

Signal detection theory describes four possible outcomes. When a signal (eg, auditory tone) is correctly perceived as present, it is a correct detection, or a "hit." When a signal is not detected even though it is present, it is a false negative, or a "miss." When a signal is absent but a perception is erroneously reported, this is a false positive, and when the signal is accurately judged absent, it is a correct rejection.

Lesson 8.2
Sensory Receptors

8.2.01 Types of Sensory Receptors

As Concept 8.1.01 describes, sensory receptors are responsible for transduction. **Sensory receptors** are specialized cells that detect stimuli in the internal (eg, blood pressure) or external (eg, light) environment and transmit this information to the nervous system. The major types of sensory receptors are mechanoreceptors, thermoreceptors, photoreceptors, and chemoreceptors.

Mechanoreceptors are sensitive to mechanical stimulation caused by pressure, vibration, or movement. For example, auditory receptors are hair cells on the basilar membrane that bend in response to sound (see Figure 8.4). This movement causes the cells to depolarize, transmitting information to the brain. Other examples of mechanoreceptors include proprioceptors, vestibular receptors, and some somatosensory receptors.

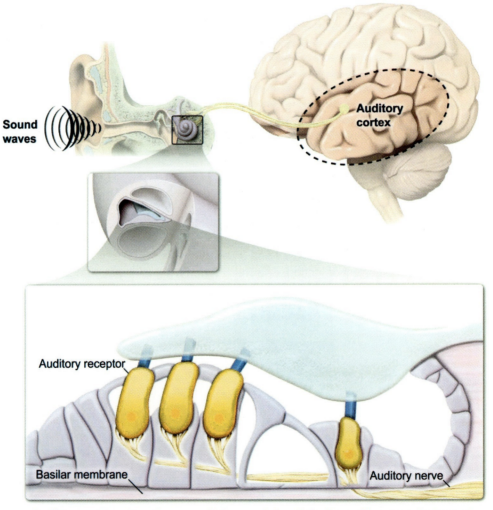

Auditory receptors transduce sound into neural signals for the auditory nerve to transmit.

Figure 8.4 Auditory transduction.

Another type of receptor that can be found in the skin is **thermoreceptors**, which are sensitive to temperature. An example of a thermoreceptor is a receptor in the skin that responds to heat.

In contrast, **photoreceptors** are sensitive to light. Rods and cones are the two main types of photoreceptors and they enable vision by converting light into neural impulses.

Chemoreceptors are sensitive to chemicals and are the type of receptors involved in taste and smell. In these chemical senses, food and odor molecules chemically activate taste and olfactory receptors, respectively.

See Table 8.1 for an overview of these sensory receptor types.

Table 8.1 Sensory receptors.

	Detects	**Stimuli**	**Example**
Mechanoreceptor	Movement	Sound waves, touch	Hair cells (ear)
Thermoreceptor	Temperature	Heat, cold	Free nerve endings (skin)
Photoreceptor	Light waves	Visible light	Rods, cones (retina)
Chemoreceptor	Chemicals	Molecules, solutes	Taste buds (tongue)

8.2.02 Sensory Pathways

Most of the pathways that carry information from the various senses have a noteworthy commonality; information from those senses travels through the **thalamus** on the way to the cortex. For example, visual information is relayed through the thalamus' lateral geniculate nucleus (LGN) on the way to the primary visual cortex.

However, in olfaction, olfactory receptor neurons transmit smell information to the olfactory bulb. The olfactory bulb relays the information to other brain areas, such as the hippocampus and the amygdala. Although sensory information from all the other senses travels to the thalamus before further brain areas, olfactory information bypasses the thalamus and is instead sent directly to other structures.

The individual sensory pathways are discussed in further detail throughout Unit 3.

Lesson 9.1

Principles of Perception

9.1.01 Top-Down and Bottom-Up Processing

Chapter 8 describes how sensory information is converted into neural signals. The process of **perception** involves integrating, organizing, and making meaning out of the data collected by the senses.

Perception guided by preexisting information or beliefs is called **top-down processing** (also called conceptually driven processing). For example, an individual misperceives a garden hose as a coiled-up snake after hearing about a snake earlier in the day.

A perceptual set describes this tendency to focus on certain details of a stimulus while overlooking other details. Culture, experiences, mood, and expectations can influence one's perceptual set. For instance, when viewing a movie about the Revolutionary War, a theater major and a history major might focus on different aspects of the movie (eg, acting versus historical inaccuracies, respectively).

In contrast, **bottom-up processing** (or stimulus-driven processing) occurs when perception is guided by the details of the sensory input. For example, an individual looks at shapes on a canvas and identifies them as animals.

Top-down and bottom-up processing are contrasted in Figure 9.1.

Figure 9.1 Top-down versus bottom-up processing.

9.1.02 Perceptual Organization

Principles of **perceptual organization** (eg, form, constancy, depth, motion) are top-down cognitive processes wherein the brain's interpretation of sensory information is guided by expectation and prior experiences. Differing expectations can cause two individuals viewing the same image to perceive different objects. The perception of form is also guided by the Gestalt principles covered in Concept 9.1.03.

The principle of **perceptual constancy** describes the tendency to perceive an object as unchanging despite slight changes to the object that occur while one is viewing it (eg, light or movement causing alterations to color, size, brightness, or shape; Figure 9.2). For instance, size constancy results in an object appearing to not change size despite changing distance (eg, a bird flying away does not appear to shrink).

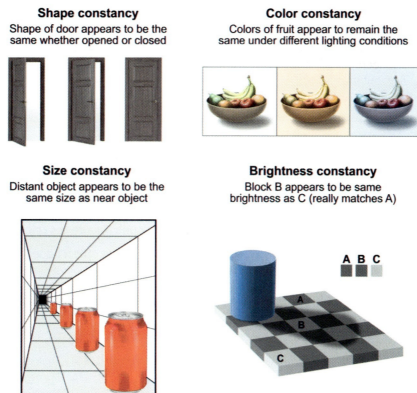

Figure 9.2 Perceptual constancy.

The ability to see in three dimensions is called **depth perception** and is enabled by the brain's interpretation of the two-dimensional information in the eye. Depth cues contribute to depth perception; monocular depth cues use information from just one eye, whereas binocular cues require both eyes.

The **monocular depth cues** (Figure 9.3) include:

- Interposition is a cue wherein an object that is partially blocked by another object is perceived as farther away.
- Relative size is a cue wherein if an individual assumes two objects are a similar size, then the one that appears smaller is perceived as farther away.
- Relative height (sometimes called height in plane) is a cue wherein an object that is lower is perceived to be closer within the visual plane than higher objects.
- Motion parallax (sometimes called relative motion) results in nearer objects appearing to pass more quickly than farther objects while the observer is moving.
- Linear perspective results in parallel lines (eg, train tracks) appearing to come together in the distance.

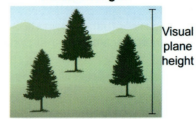

Figure 9.3 Monocular depth cues.

In contrast, the **binocular depth cues** contribute to depth perception through the integration of slightly different information from the left and right eyes. Binocular depth cues (Figure 9.4) include:

- **Retinal disparity** describes how the brain judges the distance of an object based on the difference between the visual information from each retina.
- **Convergence** describes how the eyes move together (ie, converge) to view a close object and the brain interprets the degree of convergence as an indication of the object's distance.

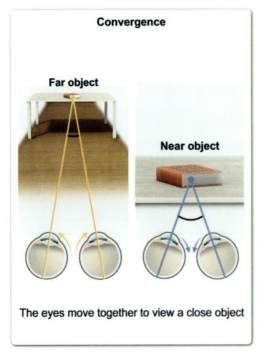

Figure 9.4 Binocular depth cues.

Visual principles of perceptual organization help explain why perceptual illusions occur. These principles allow the brain to use mental shortcuts to process visual information more quickly, which can enable optical illusions. For example, the **phi phenomenon** describes how adjacent flashing lights create the perception of motion (ie, the lights appear to move).

9.1.03 Gestalt Principles

The **Gestalt principles** of perceptual organization describe how humans perceive sensory stimuli as a whole greater than the sum of their parts. Gestalt principles apply to many types of sensory stimuli (eg, the grouping of musical tones) but are most often used to describe the perception of visual stimuli.

Examples of Gestalt principles (Figure 9.5) include:

- Subjective contours (or illusory contours) is the tendency to perceive the contours (ie, edges) of a shape even though they are not fully depicted.
- Similarity is the tendency to group together objects that share similar features (eg, shape, color).
- Continuity (or good continuation) is the tendency to perceive elements as continuing on a smooth path (eg, "X" is perceived as two crossing lines rather than two "V" shapes touching).
- Figure-ground refers to the tendency to perceive objects (ie, figures) as distinct from a background (ie, the ground).
- Closure is the tendency to perceive a whole object by filling in gaps.
- Proximity is the tendency to perceive things that are physically closer to one another as a group (eg, letters that are closer together are grouped as a word).

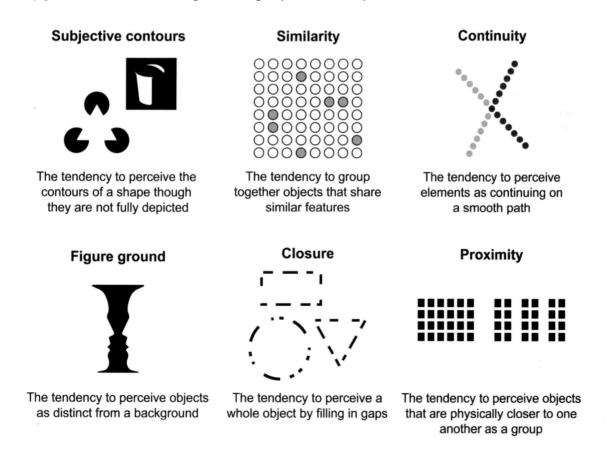

Figure 9.5 Examples of Gestalt principles.

Lesson 10.1

Eye Structure and Function

10.1.01 Eye Structure and Function

Light is an electromagnetic wave (ie, a form of electromagnetic radiation). Waves are characterized by several properties, including wavelength and amplitude. Wavelength, which describes the distance between the wave's peaks, determines light's hue (color), whereas amplitude, which describes the wave's height, determines light's intensity (contributing to brightness).

The electromagnetic spectrum arranges the different types of waves by wavelength, ranging from very long wavelength radio waves to very small wavelength gamma rays. The visible light spectrum is the part of the electromagnetic spectrum that the human eye can detect. The electromagnetic spectrum shows that different colors of visible light have different wavelengths (eg, the wavelength of green light is around 500 nm, and the wavelength of red light is around 750 nm). Visible lights often comprise different wavelengths, determining light's purity (the degree of mixture).

During the process of **vision**, light passes through several eye structures (Figure 10.1) before reaching the **retina**, the layer in the back of the eye containing photoreceptors. The outer layer of the eye is made up of two continuous segments. At the front of the eye, the cornea covers the pupil (ie, the opening in the eye's center that enables light to travel through the eye) and iris (ie, the colored part of the eye containing muscle that reacts to light's brightness by dilating or constricting the pupil). A second structure, the sclera (sometimes referred to as the white of the eye), makes up the outer layer of the rest of the eye.

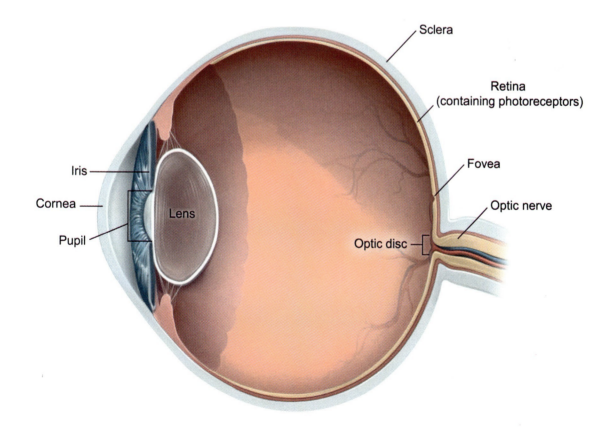

Figure 10.1 Selected structures involved in vision.

The **lens** is a transparent structure that changes shape to sharpen the image projected onto the retina. The term accommodation describes when the shape or thickness of the eye's lens changes to view objects that are close or far away.

Cells called **photoreceptors** (ie, rods and cones) in the retina convert light to neural signals. **Rods** are the receptor cells of the eye that detect movement, as well as black, white, and gray, whereas **cones** detect color. The **fovea**, a small pit in the retina with the greatest concentration of cones, has the highest level of visual acuity (ie, sharpness). Rods are more densely concentrated in the periphery of the retina and are not found in the fovea.

Signals from photoreceptors are passed to other retinal cells such as bipolar cells and then transmitted to **ganglion cells**. The axons of retinal ganglion cells, which make up the **optic nerve**, send the information to the thalamus, where it is then relayed to the visual cortex (in the brain's occipital lobe).

The optic disc is the region of the retina where the optic nerve exits the back of the eye and the vessels that supply blood to the retina enter. There are no photoreceptors in this area, resulting in a blind spot: a small gap in the visual field. The brain automatically fills in this blind spot, so it usually goes undetected.

Lesson 10.2
Visual Processing

10.2.01 Visual Processing

As Lesson 10.1 states, the optic nerve (comprising the axons of retinal ganglion cells) relays visual information from the eyes. At the optic chiasm, some axons cross over, carrying messages to the opposite brain hemisphere, and other axons continue in the same hemisphere.

From the optic chiasm, the axons (now called the optic tract) travel to the **thalamus**, where they synapse in the **lateral geniculate nucleus** (LGN), the region of the thalamus that receives visual input. From the LGN, visual information is transmitted to the **primary visual cortex** (V1) of the occipital lobe, which is the primary site of visual processing. Another area that receives visual input directly from the optic tract is the midbrain's superior colliculus, which also contributes to visual processing.

Feature Detection

David Hubel and Torsten Wiesel studied the responses of some of V1's neurons. Their work showed that cells called **feature detectors** respond only to specific visual stimuli. For example, some of these neurons fire only in response to lines at one angle, and others fire only in response to lines at other angles (Figure 10.2). Other cells in the visual system respond to more complex stimuli (eg, faces).

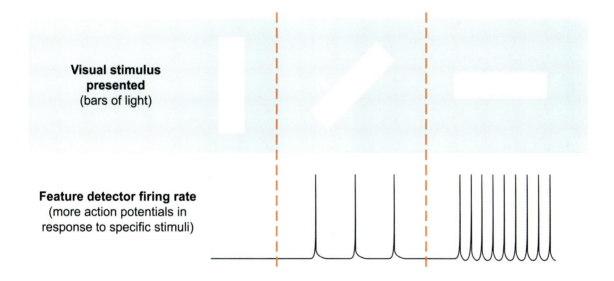

Figure 10.2 Feature detection.

Parallel Processing

Cells in V1 that are selectively responsive to orientation (ie, some feature detectors) explain the processing of just one aspect of visual stimuli. **Parallel processing** is the brain's ability to process different aspects of a stimulus simultaneously. For visual stimuli, the brain processes information about form/color and motion separately but at the same time (ie, in parallel).

LGN and V1 are organized into fairly segregated layers or regions. Information from the retina is transmitted to these regions in LGN and V1 via relatively separate pathways. The **parvo pathway**, which travels to V1 through the parvocellular layers of the LGN, includes the neurons responsible for the perception of fine object shape. Conversely, the **magno pathway**, which travels to V1 through the magnocellular layers of the LGN, includes neurons responsible for the perception of object motion.

Visual information then travels from V1 to other brain areas in two major streams. The **ventral stream**, also known as the "what" pathway, projects toward the temporal lobe and is involved in the perception of form and color. The **dorsal stream**, also known as the "where" pathway, projects toward the parietal lobe and is involved in the perception of motion.

Color Perception

The processing and perception of color begin with the cells that detect color in the retina. Two theories explain the perception of color: trichromatic theory and opponent process theory. The **trichromatic theory** states that color vision results from three types of color receptors (ie, cones) that are most sensitive to red, blue, or green.

The **opponent process theory** states that color vision results from cells that respond to color in opposing pairs: red-green and blue-yellow. An opponent process cell responds to one color of the pair (eg, red) and is inhibited by the other color (eg, green). Therefore, both colors cannot be processed at the same time.

The phenomenon of afterimages demonstrates the opponent process theory: Looking at one of the colors (eg, red) fatigues opponent process cells, resulting in seeing the other color (eg, green) on looking away.

Lesson 11.1

Ear Structure and Function

11.1.01 Ear Structure and Function

Specialized components in the outer, middle, and inner ear (Figure 11.1) contribute to **audition** (ie, hearing). The pinna, the visible part of the outer ear, funnels sound to the middle ear. Sound waves vibrate the tympanic membrane (also called the eardrum), separating the outer ear from the middle ear, which then vibrates the middle ear's ossicles. The **ossicles** are three small bones (malleus or "hammer," incus or "anvil," and stapes or "stirrup") that transmit sound vibrations to the inner ear.

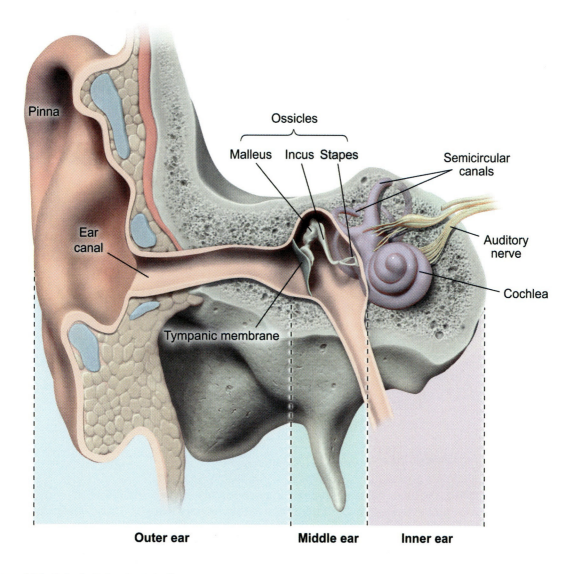

Figure 11.1 Selected structures in the ear.

The vibrations move the fluid in a spiraled inner ear structure called the **cochlea** (Figure 11.2). Inside the cochlea are three fluid-filled chambers, the middle of which houses the organ of Corti. The organ of Corti comprises the supporting cells and auditory mechanoreceptors that sit along the **basilar membrane**, a membrane at the base of the middle cochlear chamber.

The movement of endolymph (fluid in the middle cochlear chamber) results in activation of the auditory receptors. The auditory receptors are sometimes referred to as **hair cells** because on top of them sit cilia (also called stereocilia) that bend in response to movement, causing depolarization of the cell. Signals from these auditory receptors are transmitted to spiral ganglion cells, whose axons make up the auditory nerve. The auditory nerve (a branch of the vestibulocochlear cranial nerve) then relays the sound information to the brain.

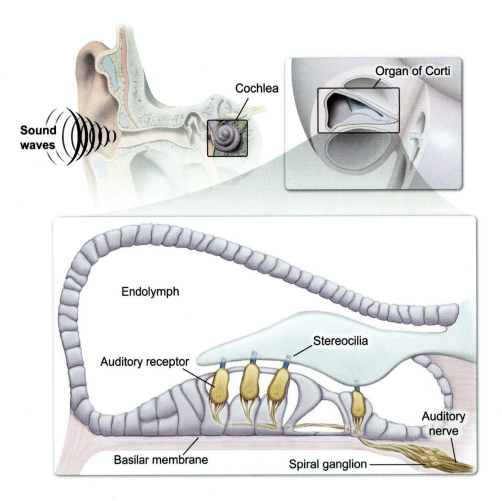

Figure 11.2 Selected structures in the cochlea.

In addition to the cochlea, the inner ear also contains the semicircular canals and the otolith organs, structures that are important in the vestibular, rather than in the auditory, sense. These vestibular structures are discussed in more detail in Lesson 12.5.

Lesson 11.2
Auditory Processing

11.2.01 Auditory Processing

After the auditory receptors transduce sound, the auditory nerve, a branch of the vestibulocochlear cranial nerve, then carries the sound information from the inner ear to the medial geniculate nucleus (MGN) of the thalamus. The MGN relays the information to the temporal lobe's primary auditory cortex, the brain's auditory processing area. Another key area that receives and processes auditory input is the midbrain's inferior colliculus.

Pitch Perception

In the process of hearing, the brain interprets sound stimuli. Pitch refers to how a sound is experienced as a high (eg, a whistle) or a low (eg, a bass drum) tone. Frequency, the length of sound waves, determines a sound's pitch.

Two theories describe pitch perception: frequency theory and place theory (Figure 11.3). **Frequency theory** states that pitch perception occurs because the frequency of a sound wave (eg, 75 waves per second) corresponds with stimulation of the auditory nerve (eg, 75 pulses per second).

In contrast, **place theory** states that pitch perception is based on the location where sounds activate receptors along the basilar membrane. Different frequencies activate receptors at specific locations.

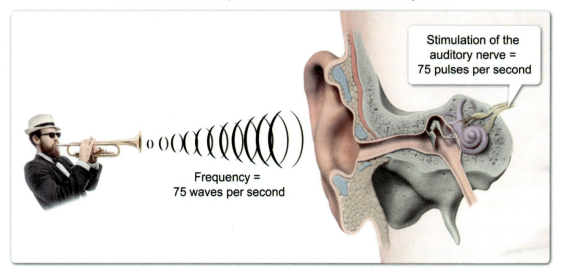

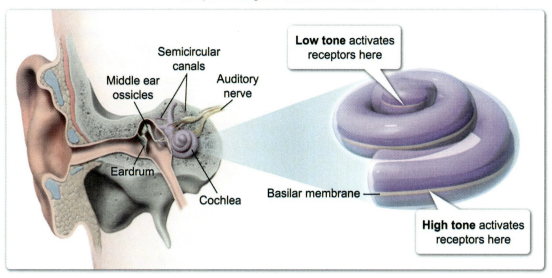

Figure 11.3 Theories of pitch perception: frequency theory versus place theory.

Auditory Localization

Auditory processing also includes auditory localization. **Auditory localization** describes how the brain perceives the source of sounds by comparing the difference between the arrival time and intensity of the sound hitting the two ears.

Lesson 12.1

Somatosensation

12.1.01 The Biological Underpinnings of Somatosensation

Somatosensation is enabled by the activation of somatosensory receptors in the skin that detect touch, pain, vibration, temperature, and movement sensations. The somatosensory receptors transduce this sensory information into neural impulses.

Different types of sensory receptors (see Concept 8.2.01) respond to different kinds of somatosensory stimuli (eg, touch, pain, temperature). Mechanoreceptors depolarize in response to mechanical stimulation caused by pressure, vibration, or movement. Thermoreceptors are sensitive to temperature (eg, depolarize in response to heat). Nociceptors are typically free nerve endings that respond to potentially harmful stimuli (ie, stimuli that can cause tissue damage), such as a sharp needle. Activation of nociceptors contributes to the perception of pain.

12.1.02 The Process of Somatosensation

In the process of somatosensation, receptors transduce somatosensory stimuli (eg, touch, temperature) into neural signals. In the context of somatosensation, afferent neurons are neurons that relay sensory information toward the spinal cord and brain. Following transduction, these sensory neurons then relay the somatosensory information from receptors through tracts in the spinal cord to the thalamus and then the primary somatosensory cortex (Figure 12.1).

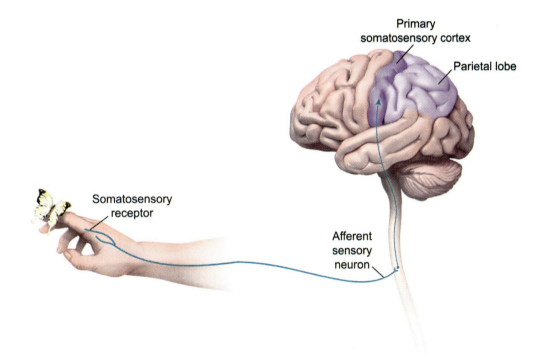

Figure 12.1 The process of somatosensation.

As Concept 4.3.02 introduces, somatosensory information is processed in the parietal lobe's somatosensory cortex. The parietal lobe is responsible for both somatosensation and proprioception (awareness of one's body position in space). The most sensitive parts of the body (eg, fingers, tongue) have the most somatosensory receptors; subsequently, these sensitive regions of the body occupy a greater area of the somatosensory cortex (see Figure 12.2).

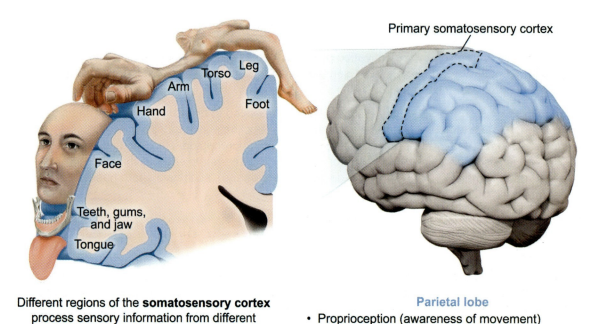

Different regions of the **somatosensory cortex** process sensory information from different parts of the body. The more sensitive parts of the body occupy a greater area of the somatosensory cortex.

Parietal lobe
- Proprioception (awareness of movement)
- Somatosensation (touch and movement sensations)

Figure 12.2 The primary somatosensory cortex.

The brain's right and left hemispheres are each specialized for certain processes; this specialization is known as hemispheric lateralization. Each hemisphere controls touch and movement on the opposite side of the body. Thus, the somatosensory cortex in the right hemisphere processes sensory information from the left side of the body and vice versa.

Pain Processing

In addition to utilizing different receptor types, the touch pathway further differs from the pain pathway. The afferent sensory neurons relaying information from nociceptors differ (eg, in size, in myelination) from other types of somatosensory afferents. Information from nociceptors also ascends in different spinal tracts (eg, in a different region of the spinal cord) though it is also relayed through the thalamus.

Pain serves as a warning of physiological damage. By drawing attention to the affected area (eg, a twisted ankle) and encouraging a change in behavior (eg, taking weight off that foot), pain can help prevent additional injury. As Concept 4.4.01 discusses, the spinal reflex enables the body to withdraw from a painful stimulus before the sensory information reaches the brain. In this reflex arc, spinal interneurons process the sensory information before it travels to the brain and directly stimulate efferent motor neurons.

One model for somatic (bodily) pain, the gate-control theory, states that pain signals to the brain are regulated by a neurological "gate" in the spinal cord. When open, the "gate" allows pain signals to reach the brain. However, massage or mental distractions can close the "gate" to block pain signals. For example, while concentrating during a game, an injured athlete feels less pain.

Lesson 12.2
Taste

12.2.01 The Biological Underpinnings of Taste

Gustation (ie, the sense of taste) involves tastants (chemicals dissolved in saliva) activating taste receptor cells in taste buds. **Taste buds**, which are found in papillae (ie, the small visible bumps on the tongue), are located across the entire tongue as well as other areas of the mouth (eg, under the tongue).

There are five primary taste qualities (also called basic tastes): sweet, salty, bitter, sour, and umami (ie, the savory flavor of glutamates such as MSG). Different kinds of papillae have slightly different sensitivities to these taste qualities, but most of the tongue is sensitive to all of the basic tastes. Each papilla contains numerous types of taste receptors, most of which respond selectively to one of the basic tastes (though some respond to multiple taste qualities). These receptor cells, which are chemoreceptors, depolarize when exposed to the corresponding tastant.

Compared to most people, supertasters have a greater number of taste buds across the tongue. They experience taste more intensely (eg, sensitivity to spicy foods) and can taste things (eg, subtle flavors) other people cannot.

The senses can be classified by the different types of sensory input (eg, odor molecules, sound waves) that are transduced. Gustation and olfaction (Lesson 12.3) constitute the **chemical senses** because transduction occurs through chemical activation of sensory receptors: food and odor molecules chemically activate taste and olfactory receptors, respectively.

12.2.02 The Process of Taste

As Concept 12.2.01 introduces, the process of taste involves tastants causing the depolarization of taste receptor cells. These cells synapse on **gustatory afferent neurons**, whose axons relay taste information to the brain through three cranial nerves. These axons travel through the brainstem to the thalamus and then the **primary gustatory cortex** for processing. The primary gustatory cortex is mostly located in the frontal lobe's insular cortex, but it also extends into other brain regions.

Sensory interaction occurs when the senses influence each other. For example, both taste and smell contribute to the perception of flavor. Tastants activate receptors found primarily on the tongue, and odors activate olfactory receptors in the nasal cavity. The nervous system sends the taste and olfactory information to the brain, where they are integrated in the perception of flavor (Figure 12.3).

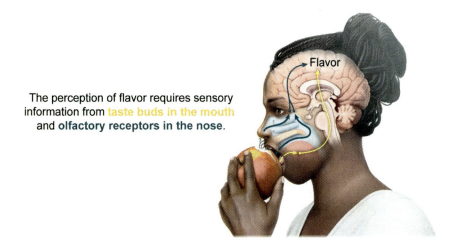

Figure 12.3 Sensory interaction in the perception of flavor.

Chapter 12: Other Senses

Lesson 12.3
Smell

12.3.01 The Biological Underpinnings of Smell

Olfaction (ie, smell) occurs when odorants (airborne odor molecules dissolved in mucus) stimulate **olfactory receptor neurons**, also called olfactory receptor cells. Odorants bind cilia on these receptors (located in the nasal cavity's olfactory epithelium), whose axons travel to the olfactory bulb (Figure 12.4). These axons collectively comprise the olfactory cranial nerve (also called the olfactory nerve).

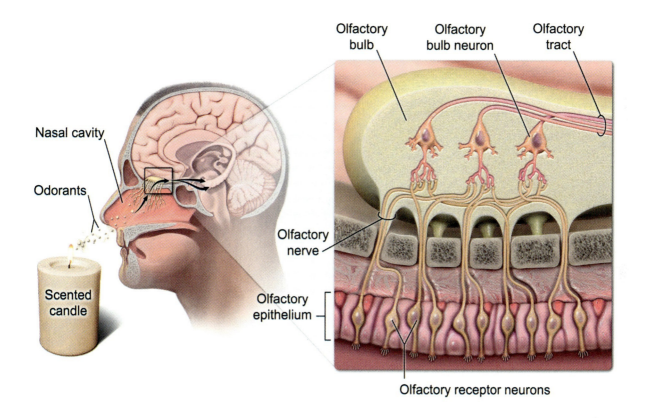

Figure 12.4 The olfactory bulb.

12.3.02 The Process of Smell

As Concept 12.3.01 describes, in the process of smell, olfactory chemoreceptors transduce olfactory information. The olfactory receptor neurons stimulate neurons within the **olfactory bulb** (see Figure 12.5). Axons from these olfactory bulb neurons comprise the **olfactory tract**. The olfactory tract relays smell information to other brain areas, such as the olfactory cortex, the hippocampus (involved in memory and learning), and the amygdala (involved in emotion).

Chapter 12: Other Senses

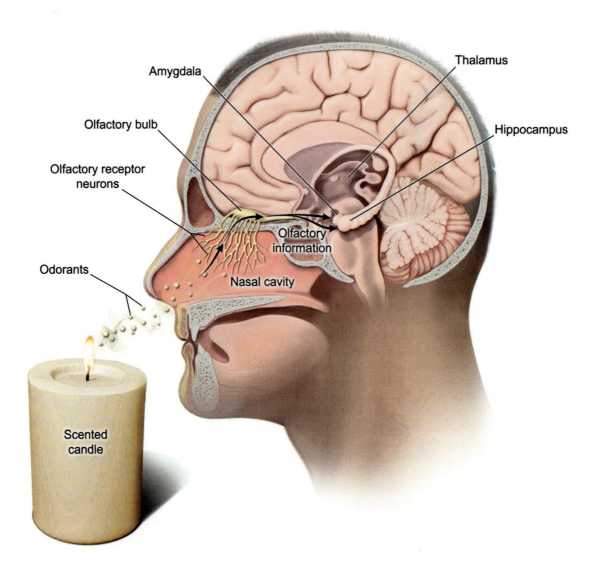

Figure 12.5 Selected structures involved in olfaction.

The olfactory pathway is distinct from the other sensory pathways in that information from all the other senses travels through the thalamus before reaching other brain areas. However, olfactory information bypasses the thalamus and is instead sent directly to cortical and temporal lobe structures.

Lesson 12.4
The Kinesthetic Sense

12.4.01 The Biological Underpinnings of the Kinesthetic Sense

The **kinesthetic sense** is awareness of body position and movement (eg, knowing where one's limbs are without looking). In the kinesthetic process (ie, proprioception), mechanoreceptors called **proprioceptors**—located in skin, joints, tendons, and muscles—respond to stretching or movement (Figure 12.6). Afferent sensory neurons then relay information from the proprioceptors to the brain.

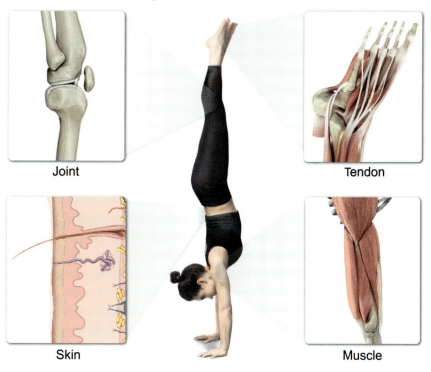

Figure 12.6 Proprioceptors.

12.4.02 The Kinesthetic Process

As Concept 12.4.01 introduces, in the kinesthetic process, proprioceptors are stimulated in response to stretching or movement. Afferent sensory neurons relay information from proprioceptors through tracts in the spinal cord to the thalamus (Figure 12.7). The proprioceptive information is then transmitted to and processed in the parietal lobe's somatosensory cortex.

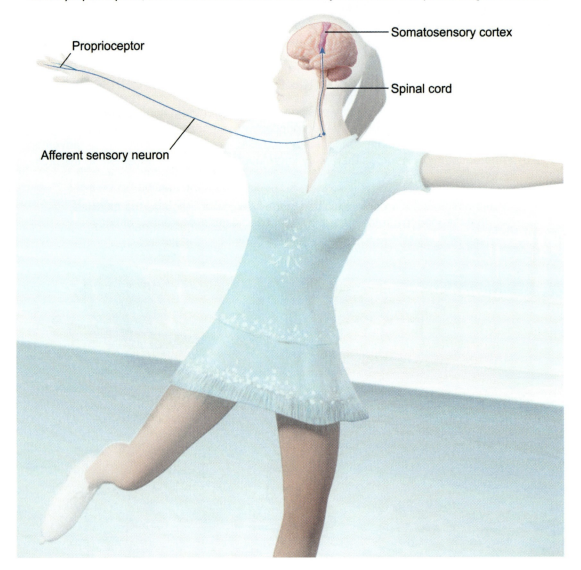

Figure 12.7 Overview of the kinesthetic process.

Lesson 12.5

Vestibular Sense

12.5.01 The Biological Underpinnings of the Vestibular Sense

The **vestibular sense** helps maintain the body's sense of balance and depends on two structures in the inner ear: the semicircular canals and the otolith organs.

The **semicircular canals** are three perpendicular fluid-filled canals (Figure 12.8). In each canal, a gelatinous structure called the cupula contains the vestibular receptor cells, hair cells. Rotation of the head moves the endolymph (fluid) in the interior of the canals, which pushes on the cupula. Displacement of the cupula bends the hair cells' cilia, causing the cells to depolarize.

When a person spins and then stops suddenly, the fluid in the semicircular canals briefly continues to move. Consequently, the sensation of spinning continues, which can make it difficult to maintain balance.

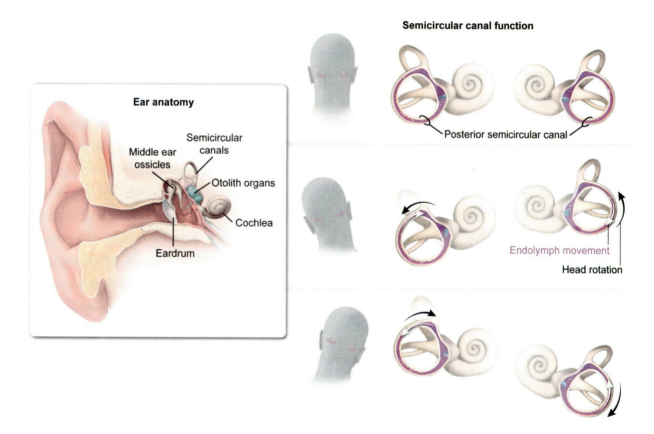

Figure 12.8 Overview of semicircular canal structure and function.

The other type of vestibular organs, also located in the inner ear, are the two **otolith organs**, the utricle and saccule (Figure 12.9). Both otolith organs contain hair cells that are located within a gelatinous membrane. On top of this membrane sit crystals that move when the head tilts. Movement of these crystals causes movement of the gelatinous membrane, which in turn bends the cilia of the hair cells, stimulating the hair cells to depolarize.

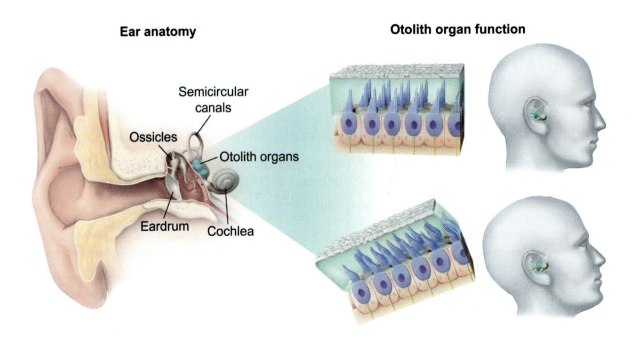

Figure 12.9 Overview of otolith organ structure and function.

12.5.02 The Vestibular Process

Together, the semicircular canals and otolith organs provide the brain with information about movement of the head. The semicircular canals detect angular acceleration: when the head rotates, hair cells in the semicircular canals are stimulated, whereas the otolith organs detect linear acceleration: when the head tilts, hair cells in the otolith organs are stimulated. The hair cells in both structures transmit the sensory information to the brain via the vestibulocochlear cranial nerve.

The **vestibulocochlear nerve** relays vestibular information directly to the cerebellum, as well as through the brainstem to the thalamus and then to the cerebral cortex. These areas coordinate posture and balance, for example, through eye movements and postural adjustments.

In addition to the contributions of the vestibular structures, maintenance of spatial orientation and balance also relies on input from the visual system and the kinesthetic sense. Visual cues provide information about the relative orientation of the body, objects, and light within one's surroundings. Proprioceptors (see Lesson 12.4) in skeletal muscles, tendons, skin, and joints provide information about the body's position and movement.

This integration of vestibular, visual, and kinesthetic inputs in order to maintain balance is an example of sensory interaction, which describes when the senses influence each other.

Lesson 13.1
States of Consciousness

13.1.01 Alertness

Consciousness is an individual's awareness of the environment and themselves. Examples of altered states of consciousness from normal alertness include sleeping and dreaming.

Multiple brain areas contribute to alertness. As Concept 4.3.01 introduces, the reticular formation (RF) is a collection of neurons that spans the brainstem (Figure 13.1). Part of the RF, the ascending **reticular activating system** (RAS) projects to the prefrontal cortex, among other cortical areas, and is important in consciousness and wakefulness. Damage to the RAS can result in sleep or even a coma.

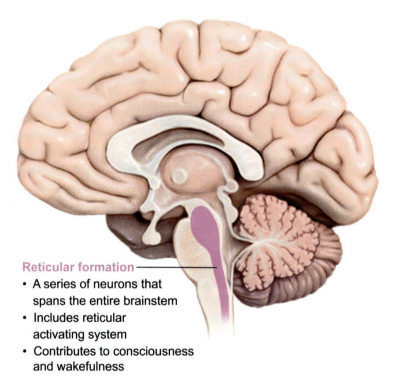

Reticular formation
- A series of neurons that spans the entire brainstem
- Includes reticular activating system
- Contributes to consciousness and wakefulness

Figure 13.1 The reticular formation.

Dissociation is the separation of certain thoughts, behaviors, or memories from normal consciousness. For example, if lost in thought while driving, a person may dissociate and arrive at a destination unable to remember the drive. In severe cases, dissociation characterizes psychological disorders involving disruption to memory and/or identity (stemming from psychological causes) (see Concept 29.1.07).

Consciousness has been described in the theories of notable historical figures in psychology. Wilhelm Wundt used introspection (ie, reporting conscious thoughts and sensations) to reveal the elements of consciousness. Another early psychologist, William James, asserted that consciousness continuously flows and is ever-changing, like a "stream of thought."

In addition, Sigmund Freud's psychoanalytic theory described three levels of consciousness: the unconscious (entirely beyond conscious awareness), the preconscious (just beneath conscious awareness), and the conscious mind. According to Freud, personality results from conflicts between the conscious and unconscious mind (eg, one's sense of right and wrong conflicting with one's impulses) (see Lesson 26.1).

13.1.02 Sleep

As Concept 13.1.01 introduces, one altered state of consciousness beyond normal wakefulness is sleep. The neural basis of sleep includes brainstem neurons, principally neurons in the pons, which play an important role in controlling sleep (Figure 13.2). Sleep is covered in more detail in Lesson 13.2.

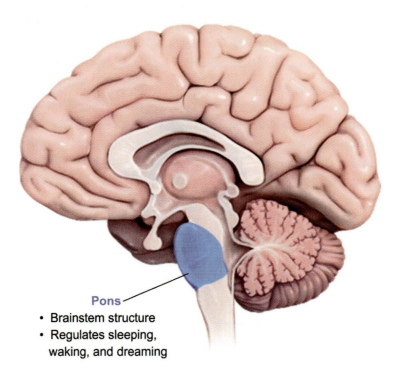

Figure 13.2 The pons.

The pons is also involved in dreaming, another altered state of consciousness. One of the theories that explains dreaming, the activation-synthesis hypothesis, states that dreams result from the brain's attempt to make sense of the random activity of the pons during sleep. The theories on dreaming are discussed further in Concept 13.2.03.

Lesson 13.2
Stages of Sleep

13.2.01 Sleep Cycles

Sleep is broadly divided into **non-rapid eye movement** (NREM) sleep (stages N1-3) and **rapid eye movement** (REM). Sleep occurs in four to six 90-minute cycles each night, and the proportion of REM increases each sleep cycle. Figure 13.3 provides an overview of the sleep stages and cycles.

The amount of time spent in each sleep stage differs over the course of a single night's sleep; a typical adult spends about 20%-25% of sleep time in REM sleep. Dreaming (see Concept 13.2.03) is most common during REM sleep but can occur in NREM stages as well.

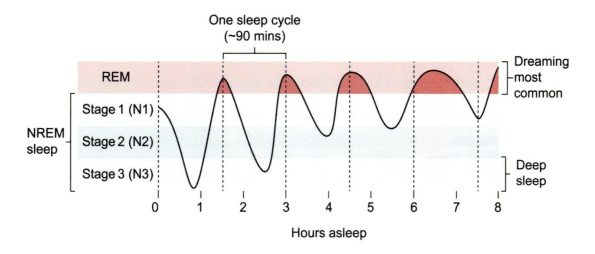

Figure 13.3 Sleep stages and cycles.

As Lesson 7.1 introduces, electroencephalography (EEG) depicts waves of brain activity reflective of the various sleep and waking states (see Figure 13.4). Alpha and beta waves are high-frequency, low-amplitude waves characteristic of waking states. **Beta waves** have the highest frequency and are typical of an awake, alert state; **alpha waves** have more regularity and are indicative of an awake, relaxed state.

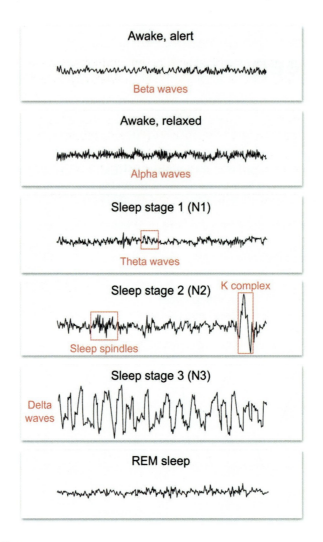

Figure 13.4 Brain wave patterns.

Stage 1 sleep (N1), or "light sleep," is marked by **theta waves**. Sudden, jerking body movements occur when an individual first falls asleep and enters the N1 stage. During NREM sleep, respiration, heart rate, and body temperature gradually decrease.

During stage 2 sleep (N2), theta waves still predominate but are interrupted by occasional sleep spindles (bursts of increases in frequency) and K-complexes (increases in wavelength). The deepest stage of NREM sleep is stage 3 (N3), slow-wave sleep. (Note: N3 reflects a consolidation of sleep stages 3 and 4 under the previous classification.) N3 is characterized by **delta waves**, which demonstrate the lowest frequency and highest amplitude observed during the sleep cycle.

In contrast, REM sleep is characterized by brain activity resembling an alert state, high-frequency, low-amplitude waves similar to beta waves. Also known as paradoxical sleep, REM sleep is marked by rapid closed-eye movements, a low body temperature, and irregular respiration and heart rates.

During typical REM sleep, muscle tone in the body is very relaxed, preventing the acting out of dreams that could result in injury. Exceptions to this near paralysis are the muscles controlling the eyes and the cardiopulmonary system.

Sleep is critical to physical and mental health; sleep deprivation can have negative effects such as fatigue, depression, difficulty concentrating, and weight gain. In the **REM rebound effect**, individuals deprived of REM sleep for even one night experience more REM sleep than usual the next night. Because they spend more time in REM, their dreams are more vivid and of longer duration. Sleep-wake disorders are discussed in Concept 13.2.04.

13.2.02 Circadian Rhythms

Circadian rhythms are cycles in physiological activity (eg, hormone release) or behavior (eg, sleeping) that occur over 24-hour intervals. One example is the sleep-wake cycle, the alternating intervals of wakefulness and sleep.

Circadian rhythms are regulated by the **suprachiasmatic nucleus** of the hypothalamus (SCN) and the pineal gland (Figure 13.5). Photoreceptors in the retina project information about light levels to the SCN. When it is dark, the SCN causes the **pineal gland** to release **melatonin**, a hormone that causes sleepiness. When light levels are high, the SCN downregulates melatonin production.

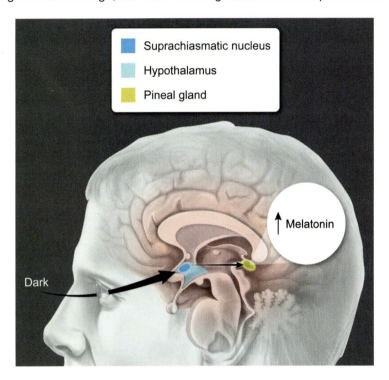

Figure 13.5 SCN regulation of melatonin release.

Most circadian rhythms, such as those involving blood pressure and core body temperature, occur in accordance with the timing of the sleep-wake cycle because they are synchronized through melatonin secretion (a light-dependent process).

13.2.03 Dreaming

As Concept 13.2.01 introduces, dreaming is most often associated with REM sleep, particularly during the REM cycles that occur closer to waking. These cycles account for the visually intense dreams remembered after waking. Although less common, dreaming can also occur during NREM sleep.

Several theories seek to explain why people dream. Concept 13.1.02 references the **activation-synthesis hypothesis**, which states that dreams result from the brain's attempt to make sense of the random activity of the pons during sleep. Another theory, the **cognitive problem-solving theory**, argues that dreams are an opportunity for the brain to solve the problems people encounter while they are awake.

In contrast, Sigmund Freud asserted that dreams are a type of wish fulfillment, in which people satisfy urges or desires they normally find unacceptable (eg, embarrassing). Freud believed that what the dream overtly seems to be about (manifest content) masks the unconscious drive or desire (latent content). For example, a student dreaming of a relaxing vacation (ie, manifest content) in which they do not have to take an anxiety-provoking exam (ie, latent content) is exhibiting wish fulfillment.

13.2.04 Sleep-Wake Disorders

Sleep disorders are conditions marked by disturbed sleep that cause distress and/or impaired functioning (Figure 13.6). Sleep-wake disorders fall broadly into two categories: dyssomnias and parasomnias. **Dyssomnias** are more common in adults and involve interference with the quality or timing of sleep, such as difficulty falling or remaining asleep, or periods of excessive sleepiness during waking hours. Examples of dyssomnias include:

- **Narcolepsy**, which involves sudden, excessive sleepiness and uncontrollable episodes of falling asleep. These narcolepsy attacks often occur while one is experiencing strong emotions (eg, while laughing with friends).
- **Insomnia**, which is defined as difficulty falling and/or staying asleep.
- **Sleep apnea**, which results in fragmented (ie, disrupted) sleep because the individual's breathing repeatedly stops, causing them to wake up briefly. This can happen hundreds of times a night. Symptoms of sleep apnea include loud snoring, choking or gasping for air, and feeling tired on awakening.

In contrast, **parasomnias** are more common in children and involve abnormal function of the nervous system during sleep, while falling asleep, or when waking up from sleep. One example is **somnambulism** (ie, sleepwalking). **Night terrors**, which is another example of a parasomnia, describes episodes of screaming, crying, or panic (ie, extreme distress) while still asleep. In contrast to nightmares, night terrors most often occur during NREM sleep and are not typically remembered.

Category	Sleep disorder	Characterized by
Dyssomnias	Narcolepsy	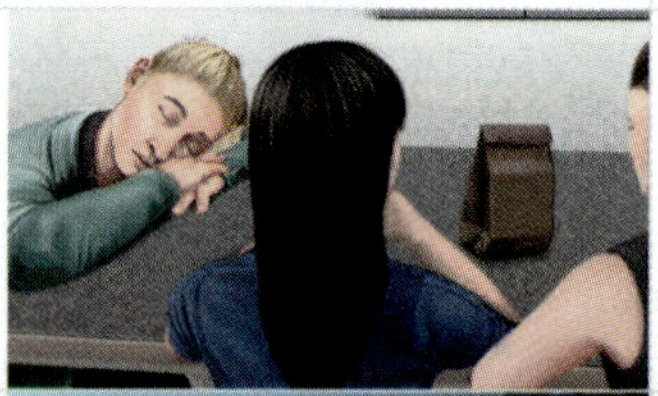 Sudden, excessive sleepiness and uncontrollable episodes of falling asleep
	Insomnia	 Difficulty falling and/or staying asleep
	Sleep apnea	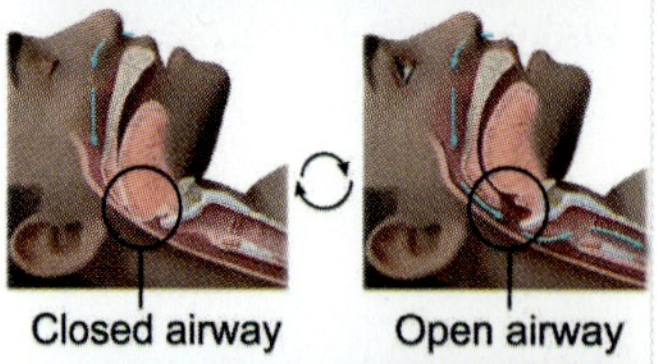 Repeated interruptions of breathing during sleep that lead to brief awakenings
Parasomnias	Somnambulism	 Sleepwalking
	Night terrors	 Episodes of screaming, crying, or panic (ie, extreme distress) while still asleep

Figure 13.6 Overview of characteristics associated with selected sleep disorders.

Lesson 14.1

Consciousness-Altering Substances

14.1.01 Consciousness-Altering Substances

Psychoactive drugs or substances affect mood, perception, thinking, and/or behavior. Categories of psychoactive drugs include stimulants, depressants, and hallucinogens (Table 14.1).

Stimulants are psychoactive drugs that increase activity in the central nervous system (CNS), resulting in an increased heart rate and feelings of well-being, energy, and alertness. Examples of stimulants include cocaine, nicotine, and caffeine. Amphetamines (eg, methamphetamine) are a class of stimulant that intensify mood and energy.

Psychoactive drugs that decrease activity in the CNS are called **depressants**. Examples of depressants include alcohol and sedatives such as benzodiazepines (eg, Valium) and barbiturates. These drugs enhance the action of GABA, the brain's principal inhibitory neurotransmitter, resulting in drowsiness, impaired coordination, and reduced behavioral inhibition.

Opioids (eg, morphine, heroin) are a class of depressants that lessen pain and produce a relaxed state by mimicking the actions of endorphins. As Concept 4.2.03 introduces, endorphins are opioids produced by the body that modulate pain, as well as contribute to elevated mood following exercise. After continued opioid use, the body produces fewer endorphins on its own, and the user experiences increased pain in the absence of the drug.

Finally, **hallucinogens** are psychoactive drugs that cause distortions in perception in the absence of sensory input (eg, seeing colorful, distorted images that are not actually present in the environment). Examples of hallucinogens include lysergic acid diethylamide (LSD), psilocybin, and mescaline. Compared to drugs in other categories, hallucinogens have the lowest risk of dependence (dependence is described in Concept 14.2.01).

Table 14.1 Psychoactive drug categories, descriptions, and examples.

Drug category	Description	Examples
Stimulants	• Increase activity in the CNS • Result in increased heart rate, feelings of well-being, energy, and alertness	• Cocaine • Nicotine • Caffeine • Amphetamines (eg, methamphetamine)
Depressants	• Slow down, inhibit, or depress the CNS • Enhance the action of GABA, the principal inhibitory neurotransmitter • Result in lowered inhibition, drowsiness, and impaired coordination	• Alcohol • Benzodiazepines (eg, Valium) • Barbiturates
	• Opioids are a subcategory of depressants that lessen pain and produce a relaxed state	• Heroin • Morphine
Hallucinogens	• Cause distortions in perception in the absence of sensory input (eg, seeing colorful, distorted images that are not actually present in the environment)	• Psilocybin • Mescaline • Lysergic acid diethylamide (LSD)

CNS = central nervous system.

Some drugs (eg, marijuana, MDMA) do not fit neatly into just one of these classifications because they have effects consistent with more than one category (eg, stimulant and hallucinogen).

Lesson 14.2
Problematic Substance Use

14.2.01 Problematic Substance Use

Psychoactive drugs or substances affect the central nervous system, resulting in changes in mood, thinking, and/or behavior. The long-term use of many psychoactive substances can cause psychological and physical dependence. **Psychological dependence** entails the belief that the substance is necessary for daily functioning (eg, a longtime smoker believes that they must smoke to feel relaxed).

Physical dependence is the body's reliance on a substance (ie, tolerance and/or withdrawal). **Tolerance** occurs when increasing amounts of a substance must be consumed to feel the same level of initial effects. Tolerance involves neurochemical changes following repeated exposure to the drug (eg, modifications to synapses or neurotransmitter receptors).

In addition, **withdrawal** occurs when physical (eg, headaches) and/or psychological (eg, anxiety) symptoms are experienced after stopping chronic substance use. Tolerance and withdrawal are contrasted in Figure 14.1.

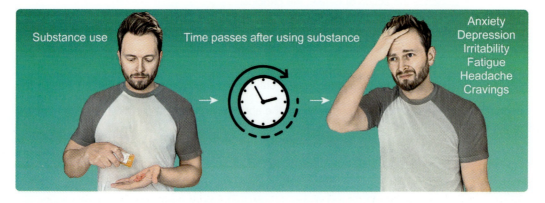

Figure 14.1 Tolerance and withdrawal.

The long-term use of many psychoactive substances can also result in substance use disorders. **Substance use disorders** are characterized by the continuation of substance use despite significant negative effects, such as a disruption to health and/or functioning (eg, relationship problems, job loss). Substance use disorders may involve symptoms such as:

- Unsuccessful attempts to cut back or stop using the substance
- Strong craving for the substance
- Significant time spent on activities related to the substance (eg, obtaining or using the substance)
- Tolerance
- Withdrawal

A region of the brain particularly involved in problematic substance use is the **mesolimbic reward pathway** (see Figure 14.2). This pathway contains the dopaminergic neurons of the midbrain's **ventral tegmental area** (VTA) (Concept 4.3.01) that project axons to the **nucleus accumbens**. The VTA also projects to different parts of the forebrain, including the lateral hypothalamus, amygdala, and prefrontal cortex.

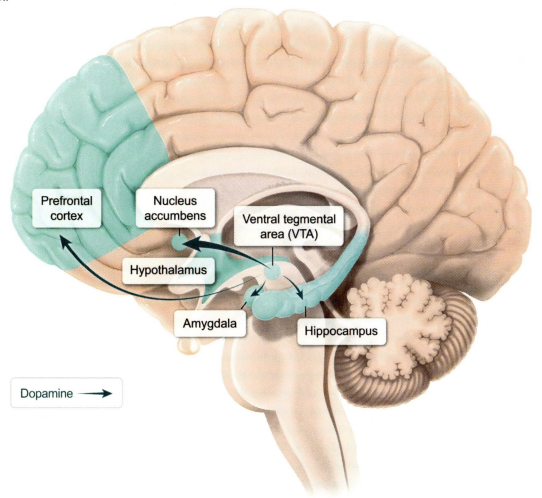

Figure 14.2 Mesolimbic reward pathway.

Rewarding stimuli (eg, food) activate this reward pathway, causing dopamine release. These areas are thought to be involved in motivation because behaviors leading to the stimulation of these neurons are reinforced. Drugs involved in substance-related disorders enhance the action of these dopaminergic neurons, and thus the use of these drugs is reinforced. Studies show that blocking the activity of dopamine disrupts the reward pathway and can decrease drug-taking and drug-seeking behavior.

END-OF-UNIT MCAT PRACTICE

Congratulations on completing **Unit 3: Sensation, Perception, and Consciousness**.

Now you are ready to dive into MCAT-level practice tests. At UWorld, we believe students will be fully prepared to ace the MCAT when they practice with high-quality questions in a realistic testing environment.

The UWorld Qbank will test you on questions that are fully representative of the AAMC MCAT syllabus. In addition, our MCAT-like questions are accompanied by in-depth explanations with exceptional visual aids that will help you better retain difficult MCAT concepts.

TO START YOUR MCAT PRACTICE, PROCEED AS FOLLOWS:

1) Sign up to purchase the UWorld MCAT Qbank
   IMPORTANT: You already have access if you purchased a bundled subscription.
2) Log in to your UWorld MCAT account
3) Access the MCAT Qbank section
4) Select this unit in the Qbank
5) Create a custom practice test

Unit 4 Learning, Memory, and Cognition

Chapter 15 Attention

15.1 Selective and Divided Attention

    15.1.01   Selective and Divided Attention

Chapter 16 Non-associative Learning

16.1 Habituation and Dishabituation, Sensitization and Desensitization

    16.1.01   Habituation and Dishabituation, Sensitization and Desensitization

Chapter 17 Associative Learning

17.1 Classical Conditioning

    17.1.01   Components in Classical Conditioning
    17.1.02   Processes in Classical Conditioning
    17.1.03   Special Types of Classical Conditioning

17.2 Operant Conditioning

    17.2.01   Reinforcement and Punishment
    17.2.02   Schedules of Reinforcement
    17.2.03   Processes in Operant Conditioning
    17.2.04   Escape and Avoidance Learning

17.3 The Cognitive Underpinnings of Associative Learning

    17.3.01   The Cognitive Underpinnings of Associative Learning

17.4 The Biological Underpinnings of Associative Learning

    17.4.01   The Biological Underpinnings of Associative Learning

Chapter 18 Observational Learning

18.1 The Process of Observational Learning

    18.1.01   The Process of Observational Learning

18.2 The Biological Underpinnings of Observational Learning

    18.2.01   The Biological Underpinnings of Observational Learning

Chapter 19 Memory

19.1 Encoding, Storage, and Retrieval

    19.1.01   Encoding
    19.1.02   Storage
    19.1.03   Retrieval

19.2 Forgetting

    19.2.01   Decay and Interference
    19.2.02   Memory Construction
    19.2.03   Aging and Memory
    19.2.04   Memory-Related Symptoms

19.3 The Biological Underpinnings of Memory

    19.3.01   Neural Plasticity
    19.3.02   Long-Term Potentiation

Chapter 20 Cognition

20.1 Cognition Across the Lifespan

    20.1.01   Piaget's Theory of Cognitive Development

20.2 Theories of Intelligence

    20.2.01   Theories of Intelligence

Chapter 21 Problem-Solving and Decision-Making

21.1 Types of Problem-Solving

    21.1.01   Types of Problem-Solving

21.2 Barriers to Effective Problem-Solving

    21.2.01   Heuristics
    21.2.02   Biases
    21.2.03   Other Barriers

Chapter 22 Language

22.1 Theories of Language Development

    22.1.01   Learning Theory
    22.1.02   Nativist Theory
    22.1.03   Interactionist Theory

22.2 Language and Cognition

    22.2.01   Language and Cognition

22.3 The Biological Underpinnings of Language and Speech

    22.3.01   The Biological Underpinnings of Language and Speech

Lesson 15.1

Selective and Divided Attention

15.1.01 Selective and Divided Attention

Only a small fraction of the sensory information from the environment is consciously processed. **Attention** refers to the cognitive processes that filter some sensory inputs to focus on others. Attention can be classified as selective or divided.

Selective attention refers to focusing on one stimulus in the environment while ignoring others.

The **cocktail party effect** is a selective attention process that occurs when an unconsciously processed stimulus triggers a person's attention, bringing it into conscious awareness (Figure 15.1). For example, when in a crowded cafeteria, an individual must tune out competing noise to focus on a conversation. But if they hear their name mentioned in the background (an unconsciously processed stimulus), their conscious attention quickly shifts to that conversation.

For example, an individual is chatting with her friend in the cafeteria.

When she hears her name (ie, stimulus) mentioned in another conversation nearby, her attention immediately shifts to the other conversation, demonstrating the cocktail party effect.

Figure 15.1 Example of the cocktail party effect.

Conversely, **divided attention** (sometimes referred to as multitasking) describes when an individual attends to more than one stimulus or task simultaneously. However, individuals generally cannot attend to multiple stimuli at the same time, so "divided attention" actually refers to rapidly switching one's attention among different stimuli or tasks.

Some tasks are easier to perform simultaneously because they can be executed via automatic processing. For example, it is easier to perform two tasks at the same time if the tasks are:

- *Dissimilar*: Driving, which requires visual attention, is easier to do while engaging in a hands-free call (auditory attention) than while texting because both texting and driving require visual attention.
- *Less difficult*: When preparing a simple, familiar dish, it is easy for a parent to simultaneously interact with their family. However, if preparing a complicated, unfamiliar dish, the parent may ask their children to play in another room.
- *Well-practiced*: A dance instructor will find it easy to simultaneously demonstrate a dance move and describe it to the class, whereas a novice dancer will find it difficult to perform the move and describe it at the same time.

In general, research shows that subjects of all ages perform poorly on divided attention tasks.

Chapter 16: Non-associative Learning

Lesson 16.1

Habituation and Dishabituation, Sensitization and Desensitization

16.1.01 Habituation and Dishabituation, Sensitization and Desensitization

One of the simplest forms of learning, **non-associative learning**, occurs when an organism changes their pattern of behavioral responding to repeated presentations of a stimulus over time.

Repeated exposure to a stimulus can result in a *decreasing* behavioral response over time, known as **habituation** (Figure 16.1). For example, a student might initially notice flickering overhead lights (ie, stimulus) at school but notice them less over time (ie, diminished response).

For example, an individual notices flickering lights.

Light flickers repeatedly over time →

The individual no longer notices flickering lights over time, demonstrating habituation.

Figure 16.1 Habituation example.

Alternatively, **dishabituation** occurs when there is a renewed (ie, increased) response to a *previously habituated* stimulus. For example, a student returns from spring break to find that the flickering overhead lights, which they had learned to ignore before the break (ie, previously habituated stimulus), are noticeable once again (ie, renewed response).

Whereas habituation refers to a diminished behavioral response after repeated exposure to a stimulus, **sensitization** occurs when repeated exposure to a stimulus produces an *increase* in a behavioral response over time. For example, after putting on an itchy sweater, an individual might scratch at it occasionally. However, over the course of the day, the individual may increasingly scratch at the sweater.

Conversely, **desensitization** occurs when the behavioral response to a *previously sensitized* stimulus decreases over time. For example, over time, an individual scratches less (ie, diminished response) at an itchy sweater that was previously unbearable (ie, previously sensitized stimulus).

These four forms of non-associative learning—habituation, dishabituation, sensitization, and desensitization—are summarized in Table 16.1.

Table 16.1 Types of non-associative learning.

	Description	Example
Habituation	Decreased response to a stimulus over time	No longer noticing that a sweater feels scratchy after wearing it for a few minutes
Dishabituation	A renewed response to a previously habituated stimulus	After taking off the sweater and then putting it back on, it feels scratchy again
Sensitization	Increased response to a stimulus over time	The sweater's scratchiness becomes more irritating until it is unbearable
Desensitization	Decreased response to a previously sensitized stimulus over time	Irritation from previously unbearable scratchiness diminishes over time

Lesson 17.1

Classical Conditioning

17.1.01 Components in Classical Conditioning

In contrast to the simpler non-associative learning, **associative learning** occurs when an organism learns the connection (ie, association) between two stimuli (as in classical conditioning) or the association between a behavior and an outcome (as in operant conditioning, see Lesson 17.2).

Classical conditioning is a type of associative learning that occurs when an organism associates a stimulus that did not previously elicit a meaningful response with a stimulus that naturally elicits a response. In the early 20th century, Ivan Pavlov, a physiologist, accidentally "discovered" this form of learning during his experiments on gustatory reflexes in dogs.

In his experiments, Pavlov noticed that the dogs, after becoming accustomed to the lab routine, started to salivate to stimuli other than their food (meat), such as a tone or bell. In other words, the dogs learned an association between a stimulus that initially produced no meaningful response, such as a bell, with a stimulus that is innately arousing, such as meat. Pavlov found that the naturally occurring response to the meat, salivation, was then produced by the bell. See Figure 17.1 for an overview of the classical conditioning process.

Several terms are key in understanding classical conditioning. The stimulus that initially did not produce a meaningful response (eg, bell) is known as a **neutral stimulus** (NS). **Unconditioned stimuli** (UCS) (eg, meat) are physiologically arousing, which means they elicit an innate (unlearned) reaction called an **unconditioned response** (UCR) (eg, salivating).

After being paired with an unconditioned stimulus, a neutral stimulus becomes a **conditioned stimulus** (CS) when it alone elicits the **conditioned response** (CR), a learned reaction (eg, salivating). The conditioned response is typically similar to, but not always exactly the same as, the unconditioned response, as is discussed in regard to conditioned taste aversions (Concept 17.1.03).

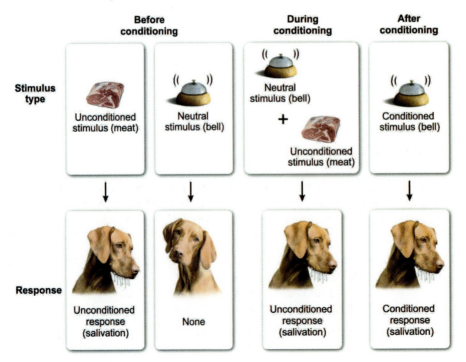

Figure 17.1 Classical conditioning overview.

17.1.02 Processes in Classical Conditioning

Acquisition, Extinction, and Spontaneous Recovery

The first phase of the classical conditioning process is known as acquisition. In **acquisition**, an association is formed between the unconditioned stimulus (eg, meat) and the neutral stimulus (eg, bell).

In many cases of classical conditioning, repeated pairings are needed for the organism to associate the neutral stimulus with the unconditioned stimulus. Across these repetitions, the association between the unconditioned stimulus and the neutral stimulus becomes stronger. During this phase, the previously neutral stimulus will eventually take on the properties of the unconditioned stimulus and elicit the now-conditioned response (eg, salivation). The neutral stimulus becomes known as the conditioned stimulus at this point.

A graph of the strength of the conditioned response across all phases of the classical conditioning process is shown in Figure 17.2.

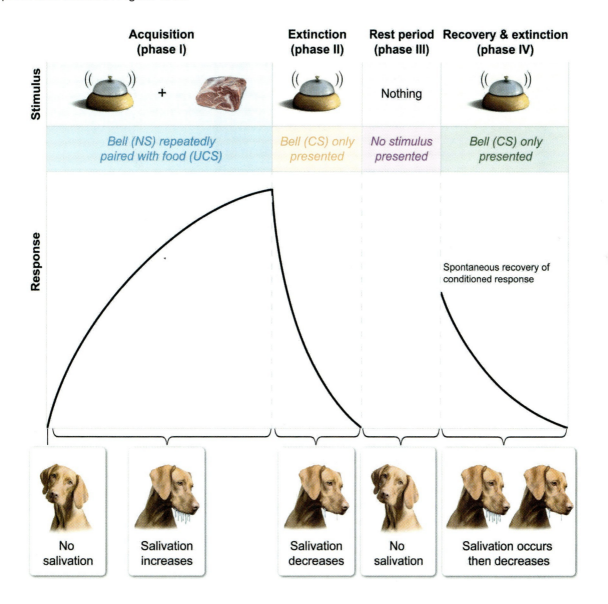

Figure 17.2 Graph showing the strength of the conditioned response.

The next phase of classical conditioning involves presentations of the conditioned stimulus (eg, bell) alone in the absence of the unconditioned stimulus (eg, meat). At first, in the absence of the unconditioned stimulus, the conditioned stimulus will initially elicit a strong conditioned response (eg, salivation). However, over repeated presentations, the conditioned response will decrease and eventually cease altogether, a process known as **extinction**.

The graph in Figure 17.2 depicts a "pause" or rest period following extinction in which the conditioned stimulus is not presented. If the conditioned stimulus is reinstated after this, it can lead to **spontaneous recovery**, in which the organism responds to the conditioned stimulus with the conditioned response once again.

Discrimination and Generalization

In classical conditioning, **discrimination** (also called stimulus discrimination) occurs when an organism responds to certain conditioned stimuli but ignores similar stimuli. For example, discrimination is demonstrated by a dog who has been conditioned to salivate in response to a bell salivating only to the sound of that specific bell, not in response to other similar bell tone sounds, such as a cell phone alert.

Similarly, **generalization** (also called stimulus generalization) occurs when a stimulus similar to the original stimulus evokes the same, conditioned response. For example, a dog who has learned to salivate in response to a specific bell tone would demonstrate generalization when it salivates in response to a similar-sounding tone, such as a cell phone alert.

See Concept 17.2.03 for discrimination and generalization in operant conditioning.

17.1.03 Special Types of Classical Conditioning

Conditioned Taste Aversions

A **conditioned taste aversion** (also called learned taste aversion) is a specific and powerful type of classical conditioning that occurs after an organism becomes ill following the consumption of a food or beverage (see Figure 17.3).

For example, if an individual eats a donut (neutral stimulus) and then gets a stomach virus (unconditioned stimulus) and becomes sick (unconditioned response), the individual will avoid (conditioned response) donuts (conditioned stimulus) in the future. A learned taste aversion can cause an individual to feel sick when even just thinking about a particular food, and the aversion may generalize to related types of foods (eg, all types of cake).

Conditioned taste aversions almost always link illness with foods (or smells), which is thought to be an evolutionary adaptation. Conditioned taste aversions occur because of **biological preparedness**, the tendency to readily learn associations that promote survival.

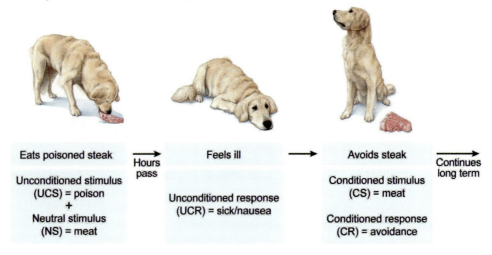

Figure 17.3 Conditioned taste aversion example overview.

Conditioned taste aversions possess several characteristics that render them a unique form of classical conditioning. Unlike typical classical conditioning, which usually requires two stimuli to be paired together repeatedly before the organism learns to associate the two, a conditioned taste aversion develops after just one pairing. In other words, an organism needs to become ill *only once* to associate the food or beverage consumed with the illness.

Furthermore, conditioned taste aversions differ from typical cases of classical conditioning in the time frame needed between the presentation of the neutral stimulus and unconditioned stimulus for the organism to form an association. Whereas typical classical conditioning requires the two stimuli to both be presented within a very short time frame for the organism to learn to associate them (with the presentation of the neutral stimulus ideally occurring just slightly prior to the unconditioned stimulus), taste aversions can be learned despite *hours* passing between the consumption of a food and subsequent illness.

Whereas typical classical conditioning rapidly extinguishes when the two stimuli are no longer paired, conditioned taste aversions have long durations. In other words, after becoming ill, the organism may never consume the associated food again.

Last, whatever was consumed prior to illness can become associated with the illness (even if it did not cause the illness) and is avoided afterward. For example, if an individual experiences nausea and vomiting in the afternoon because they contracted the flu, they may develop a conditioned taste aversion to any of the foods or beverages they consumed earlier that day, even though the food and beverages did not cause the illness.

Classically Conditioned Phobias

John Watson, often considered the founder of behaviorism, took inspiration from the work of Pavlov to study classically conditioned emotions in humans. In an experiment wrought with ethical concerns, John Watson and colleagues classically conditioned an infant known as "Little Albert" to fear white rats. See Figure 17.4 for an overview of this study.

In the Little Albert experiment, a white rat (NS) was paired with a loud noise (UCS) that caused fear (UCR) in Little Albert. After the loud noise was paired with the white rat, the white rat alone (CS) provoked fear (CR). Furthermore, Little Albert's fear of the white rat generalized to other fuzzy, white stimuli such as white rabbits and even white beards.

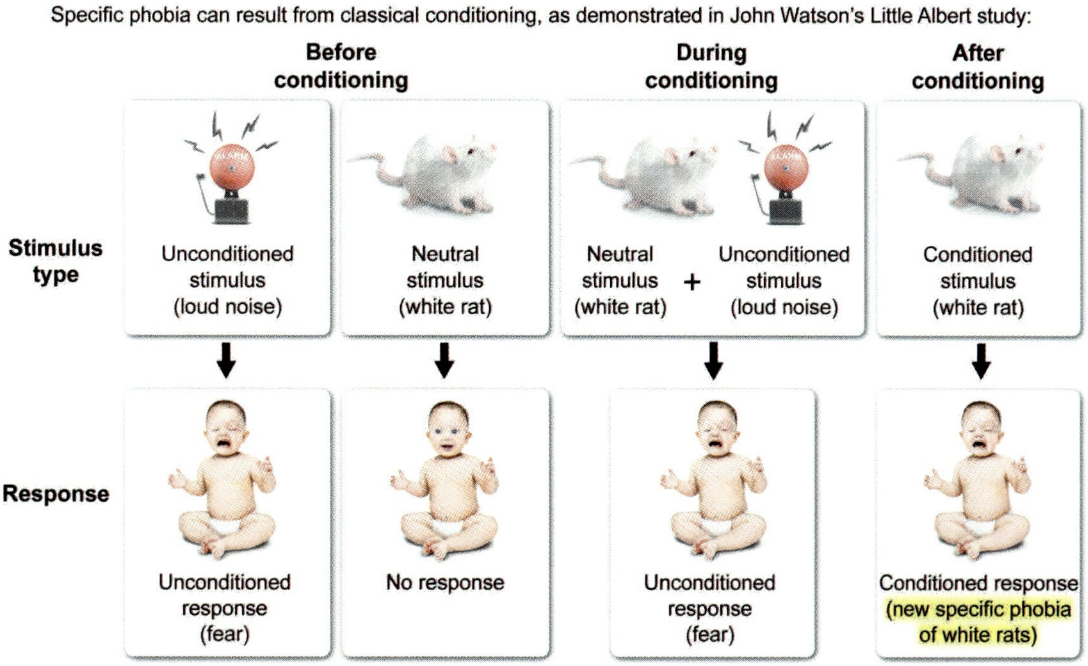

Figure 17.4 John Watson's Little Albert experiment.

The results of the Little Albert experiment demonstrated that fear can be classically conditioned. This type of fear response is considered analogous to the psychological disorder known as specific phobia. Specific phobia, covered in Concept 29.1.01, is an anxiety disorder characterized by excessive, irrational fear of a specific situation or animal/object. Some specific phobias are hypothesized to result from the classical conditioning of fear through pairing a negative experience (eg, nearly drowning) with a specific object (eg, pool) or situation (eg, swimming).

These same classical conditioning principles can be used in behavioral therapy to decrease a conditioned fear response (see Lesson 31.1).

Lesson 17.2

Operant Conditioning

17.2.01 Reinforcement and Punishment

In the 1930s, another type of associative learning, operant conditioning, was described by B.F. Skinner. **Operant conditioning** occurs when an organism associates a behavior with a consequence. In operant conditioning, the likelihood of an organism repeating the behavior is influenced by the outcome of that behavior (ie, reward or punishment). For example, when a rat receives a food pellet (ie, reward) after pushing a lever, the rat is more likely to push the lever again.

Behaviors *increase* due to reinforcement: positive reinforcement occurs when a desirable stimulus is applied, leading to an increased likelihood of behavior. For example, an individual compliments her boyfriend and smiles (ie, applies a desirable stimulus) after he cooks her dinner (ie, a behavior), which encourages him to cook more often.

Similarly, negative reinforcement occurs when an undesirable stimulus is withdrawn, leading to an increased likelihood of behavior. For example, an individual's car stops making an annoying beeping sound (ie, removes an undesirable stimulus) after she buckles her seatbelt (ie, a behavior), which increases her buckling behavior.

Behaviors *decrease* due to punishment: positive punishment occurs when an undesirable stimulus is applied, resulting in a decreased likelihood of behavior. For example, an individual yells at her puppy (ie, applies an undesirable consequence) for jumping on guests (ie, a behavior), which decreases her puppy's jumping.

Additionally, negative punishment occurs when a desirable stimulus is withdrawn, resulting in a decreased likelihood of behavior. For example, a parent takes away a child's video games (ie, removes a desirable stimulus) in response to the child acting out (ie, a behavior), which leads to the child acting out less in the future.

See Figure 17.5 for a summary of positive and negative reinforcement and punishment.

Figure 17.5 Reinforcement and punishment in operant conditioning.

17.2.02 Schedules of Reinforcement

Schedules of reinforcement (also called reinforcement schedules) are used in operant conditioning to train and/or maintain learned behaviors. Schedules of reinforcement reward (ie, reinforce) an organism either based on the frequency of responses (ie, *ratio*), or time (ie, *interval*). The schedules are either unchanging (ie, *fixed*) and are therefore predictable or based on an average (ie, *variable*) and are therefore unpredictable.

Examples of reinforcement schedules include:

- **Fixed-ratio** schedules, which provide rewards after a predictable number of responses (eg, receiving a free sandwich after every 10 purchases).
- Continuous schedules, which reward every response (eg, petting a dog every time it puts its chin on its owner's lap), are a type of fixed-ratio schedule. Continuous schedules are often used when initially training a behavior.
- Variable-ratio schedules, which provide rewards after an unpredictable number of responses (eg, only occasionally winning money by playing slot machines), are the type of reinforcement schedule most resistant to extinction (see Concept 17.2.03 for information on extinction).
- **Fixed-interval** schedules, which provide rewards after a predictable amount of time regardless of how many behaviors have occurred (eg, being paid a biweekly salary).
- **Variable-interval** schedules, which provide rewards after an unpredictable amount of time regardless of how many behaviors have occurred (eg, checking the front door frequently even though this has no impact on when a package will be delivered).

Aside from the continuous schedule, which rewards every response, the other types of reinforcement schedules listed are partial reinforcement schedules, which means they do not reward based on every response or a specific unit of time. See Figure 17.6 for a summary of these schedules.

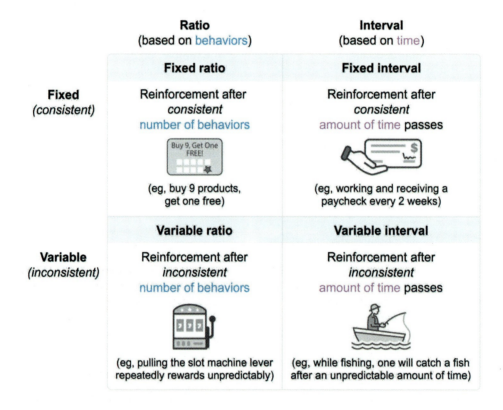

Figure 17.6 Types of partial reinforcement schedules.

Each reinforcement schedule produces characteristic behavioral response patterns (see Figure 17.7). The ratio schedules, which provide reinforcement after a consistent (fixed) or inconsistent (variable) number of behavioral responses, both produce rapid response rates. The interval schedules, which provide reinforcement after a consistent (fixed) or inconsistent (variable) amount of time, both produce slower response rates.

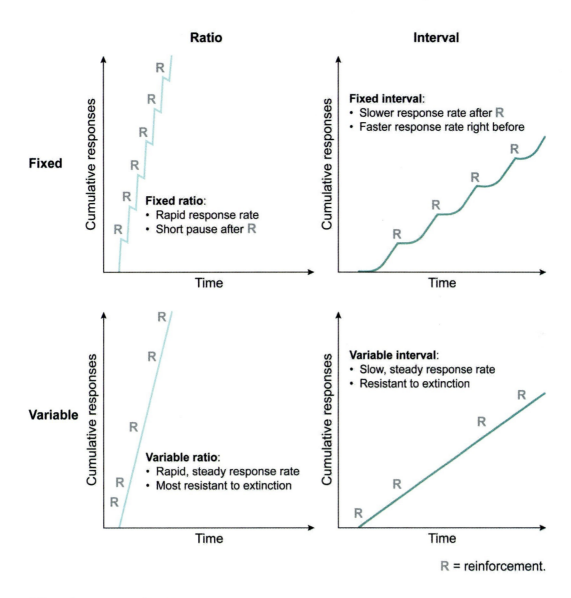

Figure 17.7 Reinforcement schedule responses.

Furthermore, although the term *reinforcement schedule* and the examples above use *reinforcement*, these schedules can also be applied via punishment. For example, if a child is grounded at each instance of being rude to his parents, this is punishment on a *fixed-ratio* schedule.

17.2.03 Processes in Operant Conditioning

Several additional terms are critical to understanding operant conditioning processes.

One way to train a new behavior is through shaping (see Figure 17.8). **Shaping** is a technique used in operant conditioning in which successive approximations (ie, behaviors that progressively resemble the desired behavior) are reinforced (ie, rewarded). For example, when teaching her dog to go into a kennel (desired behavior), an individual reinforces the dog first for sitting near the kennel, second for touching the kennel, third for putting a paw into the kennel, and so on until the dog goes completely into the kennel.

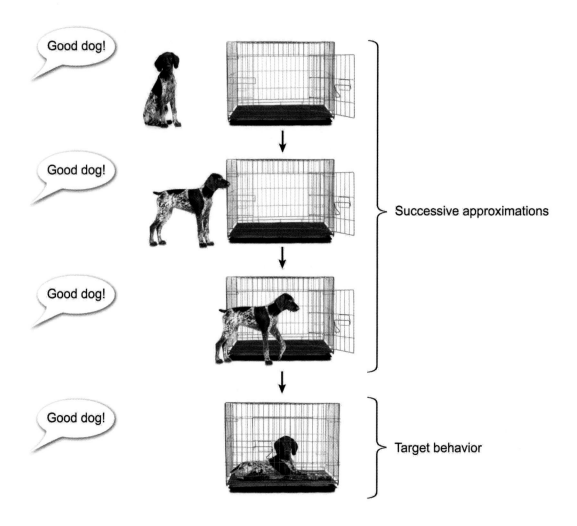

Figure 17.8 Shaping example.

In contrast, **extinction** occurs in operant conditioning when a behavior decreases or stops because it is no longer reinforced (see Concept 17.1.02 for extinction in classical conditioning). Consider the example of an individual who often bakes cookies for her friend. At first, the friend always remembers to thank the cookie baker for the cookies and praises her. However, over time, the friend forgets to praise the baker, which eventually causes the baker to stop baking the cookies, resulting in extinction.

Concept 17.1.02 on classical conditioning discusses stimulus discrimination (also called discrimination) and stimulus generalization (also called generalization) in the context of classically conditioned stimuli. In operant conditioning, **generalization** occurs when a stimulus similar to the original evokes the same response. For example, a young child has been reinforced for saying "flowers" when she sees colorful flowers. When the child sees colorful leaves (ie, new stimulus) that are similar to the original stimulus (ie, colorful flowers), she points and says "flowers" (ie, gives the same response).

Conversely, **discrimination** occurs in operant conditioning when an organism responds to certain stimuli but ignores similar stimuli. For example, a dog lies down (ie, responds) *only* when she hears her owner say the word "down" but ignores similar stimuli, such as her owner saying the word "done."

Lastly, reinforcers can be classified as primary or secondary (Figure 17.9). A **primary reinforcer** is a stimulus that is innately rewarding to an organism, such as food; an organism does not need to learn that a primary reinforcer is rewarding. However, a **secondary reinforcer** (also known as a conditioned reinforcer or a conditional reinforcer) (eg, money) is a stimulus that has been associated with a primary reinforcer. Secondary reinforcers can sometimes be used to acquire a primary reinforcer (eg, money can buy food).

Chapter 17: Associative Learning

Figure 17.9 Primary versus secondary reinforcer example.

17.2.04 Escape and Avoidance Learning

Concept 17.2.01 on operant conditioning discusses negative reinforcement, which is the withdrawal of an unpleasant stimulus following a behavior, resulting in an increase in the likelihood that the behavior will occur again. Negative reinforcement can lead to escape learning and/or avoidance learning (see Figure 17.10).

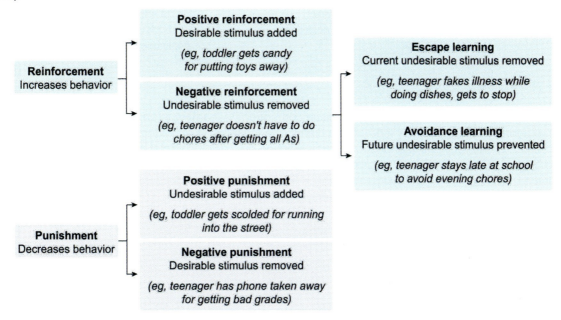

Figure 17.10 Escape learning and avoidance learning.

Escape learning occurs when an organism learns how to terminate an ongoing unpleasant stimulus (eg, a dog jumps over a partition to flee from or stop a continuous electric shock). Escape learning becomes **avoidance learning** when an organism prevents coming into contact with an unpleasant stimulus (eg, a dog jumps over a partition to avoid the electric shock before it occurs).

Lesson 17.3
The Cognitive Underpinnings of Associative Learning

17.3.01 The Cognitive Underpinnings of Associative Learning

Cognition plays an important role in associative learning. In classical conditioning, an **expectancy** refers to an organism's awareness that the unconditioned stimulus (eg, shock) will likely follow the neutral stimulus (eg, tone). For example, if a tone comes before a shock on several occasions, a rat will begin to expect the upcoming shock upon hearing the tone and consequently display distress (Figure 17.11).

Research has demonstrated that animals can learn about the degree to which the tone predicts the shock (ie, its predictive value); a tone that precedes a shock *every* time will elicit a stronger response than a tone that precedes a shock only *some* of the time.

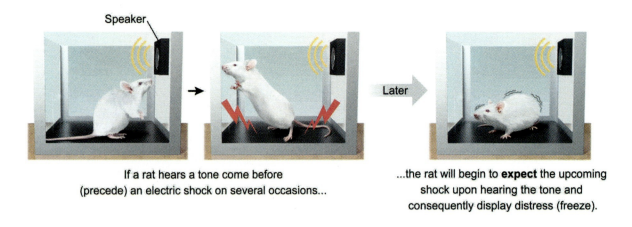

Figure 17.11 Example of expectancy in classical conditioning.

Studies have also shown that there is a cognitive component to learning in an operant conditioning task. In one experiment, rats passively learned the layout of a maze while exploring in the absence of reinforcement. The rats later demonstrated their learning by quickly completing the maze for a food reward (Figure 17.12). This study suggests that, even in the absence of a reward, it is possible for organisms to develop **cognitive maps** (ie, mental images of physical space) that can be used when needed, such as when a food reward is suddenly presented.

Chapter 17: Associative Learning

Phase 1

For 10 days, 2 groups of rats were tested in a maze.
Group A was given a food reward for completing the maze; Group B was not.

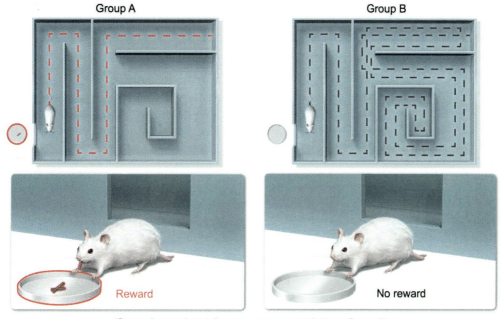

Group A completed the maze more quickly than Group B.

Phase 2

On the 11th day, Group A was no longer given a food reward,
whereas Group B had a food reward added.

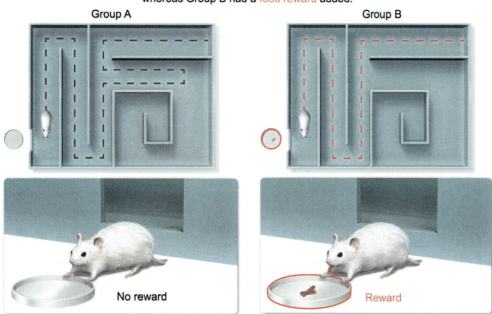

Group B completed the maze more quickly than Group A, providing evidence
for the existence of *cognitive maps* (ie, mental images of physical space).

Figure 17.12 Experiment supporting cognitive maps.

Lesson 17.4

The Biological Underpinnings of Associative Learning

17.4.01 The Biological Underpinnings of Associative Learning

An organism's biology can hinder or facilitate associative learning. Classical conditioning (Lesson 17.1) is aided by **biological preparedness**, the tendency of people or animals to readily learn associations that promote survival (eg, taste aversions) (Figure 17.13). For example, animals more readily develop specific phobias to situations (eg, confined spaces) or animals that are potentially harmful (eg, snakes).

When monkeys are exposed to a stimulus (ie, toy) that resembles...

...a harmless object (ie, flowers)...   ...a potentially harmful animal (ie, snakes)...

...monkeys do not demonstrate fear.   ...monkeys demonstrate fear.

Monkeys more readily develop phobias of toys that look like snakes (ie, resemble a potentially harmful animal) as compared to toys that look like harmless objects (eg, flowers).

Figure 17.13 Example of biological preparedness in monkeys.

In contrast, operant conditioning (Lesson 17.2) can be limited by biological factors; one example is instinctive (or instinctual) drift. An instinct is an innate, fixed pattern of behavior that is more complex than a reflex, which is a simple response to a stimulus (eg, jerking one's hand away from a hot stove).

Instincts are not based on prior experience or learning. For example, newly hatched sea turtles instinctively know to move toward the ocean and swim.

Instinctive drift describes when an animal's innate behaviors overshadow a learned behavior. Animals trained using operant conditioning (whereby a desired behavior is reinforced) will often revert to innate behaviors even when reinforcement is provided (Figure 17.14).

For example, researchers successfully used food rewards to train pigs to pick up wooden coins and deposit them into a piggy bank. Over time, the pigs began dropping the coins before reaching the piggy bank and pushing them along the ground with their snouts, an innate behavior known as rooting.

Animal **learns new behavior**
through operant conditioning
(eg, pig taught to put coins into a piggy bank)

Instinctive drift: Learned behavior
replaced by innate behavior
(eg, trained pig starts dropping coins and rooting)

Figure 17.14 Example of instinctive drift in pigs.

Lesson 18.1

The Process of Observational Learning

18.1.01 The Process of Observational Learning

Observational learning, also known as social learning, occurs when an observer imitates a behavior that someone else has modeled (Figure 18.1). For example, people can learn new behaviors by watching others demonstrate those behaviors, such as when a medical student (observer) sees an experienced surgeon perform a new technique (model) and so they mimic that technique (imitation).

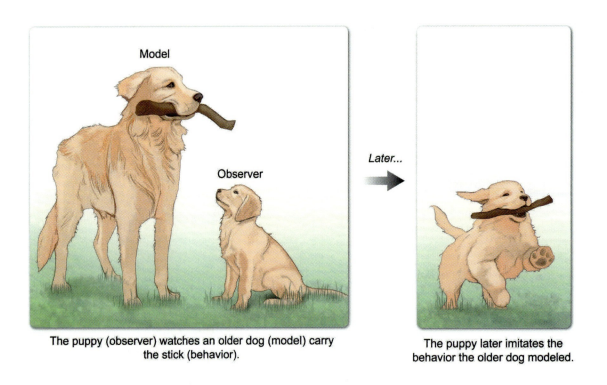

The puppy (observer) watches an older dog (model) carry the stick (behavior).

The puppy later imitates the behavior the older dog modeled.

Figure 18.1 Example of observational learning.

Albert Bandura, an early researcher in observational learning, conducted the "Bobo doll" experiments. These influential studies demonstrated that children who observed others acting aggressively toward a "Bobo doll" were more likely to display aggression toward the doll themselves.

Lesson 18.2

The Biological Underpinnings of Observational Learning

18.2.01 The Biological Underpinnings of Observational Learning

Specialized neurons called **mirror neurons** fire when organisms engage in a particular behavior and when they observe that behavior in others. Mirror neurons are found in multiple brain regions, including the frontal lobe's (Figure 18.2) motor cortex, an area of the brain responsible for planning and initiating voluntary movement.

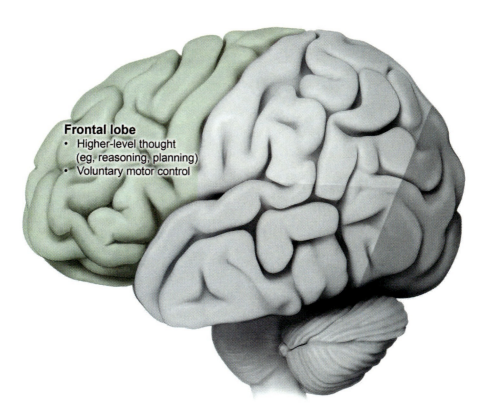

Figure 18.2 The frontal lobe.

Observational learning (see Lesson 18.1) occurs when an observer imitates a behavior that someone else has modeled. Research shows that observational learning involves the mirror neuron system. Mirror neurons play a role in imitation (the copying of another's behavior) because they fire when an organism watches or replicates a behavior.

Lesson 19.1

Encoding, Storage, and Retrieval

19.1.01 Encoding

Memory involves **encoding**, the transfer of information into memory; storage (retaining the information); and retrieval (accessing the information) (Figure 19.1). Some information is processed automatically with little effort (eg, registering the characters on a license plate as letters or numbers), but to encode information, attention and effortful processing are often required (eg, encoding a specific license plate takes effort).

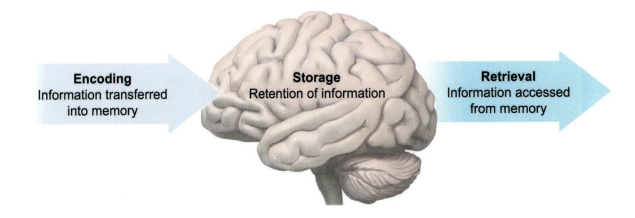

Figure 19.1 Memory processes: encoding, storage, retrieval.

Levels of processing is a concept that describes how information processed at a deeper level is encoded and retrieved (ie, remembered) better than information processed on a shallower level. One encoding strategy that can enhance memory is **elaboration**, in which new information is meaningfully associated with previously known information. This effortful, deep processing tends to result in more connections to the new material, improving learning.

For example, consider a student studying the meaning of the words "ventral" and "dorsal." While studying the word ventral, she focuses only on the appearance of the word and notices that it starts with a "v." In contrast, while studying the meaning of the word dorsal (ie, "to the back"), she uses elaboration by meaningfully associating this information with her favorite animal, dolphins, which have dorsal fins on their backs. As a result, this student has better recall of the meaning of the word dorsal as compared to ventral because she processed information about the word dorsal at a deeper level (Figure 19.2).

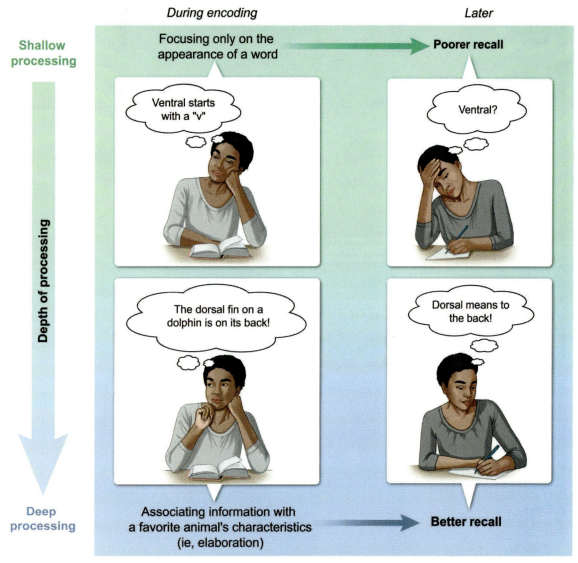

Figure 19.2 Elaboration example.

Similarly, the **self-reference effect** states that information that is personally relevant (ie, linked to oneself) is easier to remember because personally relevant information is meaningful and so is processed at a deeper level. For example, an individual who has difficulty remembering passwords changes their email password to the date of their wedding, making the password personally relevant, which leads to easier retrieval (Figure 19.3).

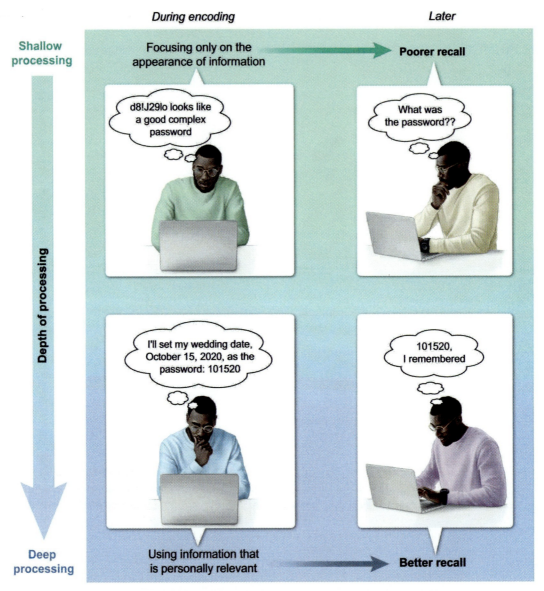

The individual has better recall of a password that is his wedding date (ie, personally relevant) because he processed the information at a deeper level, demonstrating the self-reference effect.

Figure 19.3 Self-reference effect example.

Additionally, **mnemonics** are strategies (eg, songs, acronyms) that aid memory encoding and retrieval. For example, an individual uses an acronym to help her remember the colors of the rainbow (Figure 19.4).

For example, a student uses an acronym (ie, a mnemonic) to help her remember the colors of the rainbow.

Figure 19.4 Mnemonic example.

19.1.02 Storage

As Concept 19.1.01 introduces, memory involves three steps: encoding, storage, and retrieval. **Storage** refers to retaining the encoded information. The three types of memory, sensory memory, short-term memory, and long-term memory, hold information for varying amounts of time (Figure 19.5). **Sensory memory** first briefly and temporarily stores information from the environment (eg, sights, sounds). Iconic memory is sensory memory of visual information, whereas echoic memory is sensory memory for auditory information.

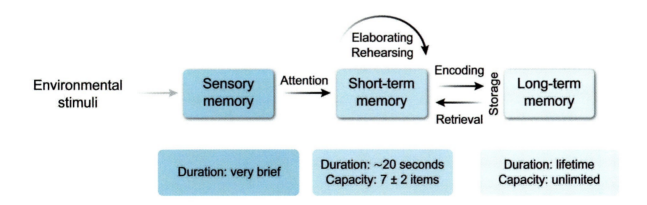

Figure 19.5 Three-stage model of memory.

Short-term memory then stores pieces of information from sensory memory. Short-term memory has a short duration (about 20 seconds) and a storage capacity of about seven items (plus or minus two). Maintenance rehearsal (mentally repeating something over and over) can prolong the duration of short-term memory. Chunking describes the strategy of grouping items into clusters that are more easily held in short-term memory (eg, grouping the digits of a phone number into three chunks rather than a long series of individual digits).

Information not transferred from short-term to long-term memory is lost. **Long-term memory** has a large capacity and a long duration (memories can be stored permanently) and comprises two branches: implicit memory and explicit memory.

Implicit memory (also called nondeclarative memory) is memory for things that cannot be consciously recalled, such as skills, tasks, emotions, and reflexes. Implicit memory includes procedural memory, which is memory for well-learned motor skills (eg, riding a bicycle). Another example of implicit memory is emotional/reflexive memory, which is memory for associations between stimuli (eg, salty ocean air triggers pleasant emotions from childhood beach vacations).

In contrast, **explicit memory** (also called declarative memory) is memory for facts and events that can be consciously or intentionally recalled. Explicit memory includes episodic memory, which is the memory for personal experiences (eg, what one ate for dinner last night), and semantic memory, which includes knowledge about facts (eg, Austin is the state capital of Texas) and language (eg, "their" and "there" are not synonymous).

Semantic long-term memory appears to be organized as a network of interconnected nodes containing factual concepts (eg, colors, objects). The organization and relationship between nodes (how linked or connected they are in memory) is unique to each individual because of the personal meaning associated with each node. For example, an individual with an uncle who is a firefighter may think of "uncle" when viewing a fire engine, which would not occur for most people.

The **spreading activation model** suggests that when a node in the semantic network is activated (eg, viewing a picture of a toy fire engine), nodes directly connected to that node (eg, siren, alarm) are activated as well, which is known as priming.

19.1.03 Retrieval

As Concept 19.1.01 introduces, the final step involved in memory is retrieval. **Retrieval** refers to accessing the encoded information from storage. There are three types of memory retrieval processes: recall, recognition, and relearning.

- Recall is the retrieval of information previously encoded.
- Recognition involves the correct identification of information that one has been exposed to.
- Relearning involves re-encoding information that was previously learned but forgotten. Typically, relearning happens much faster than learning something for the first time.

Memory retrieval can be aided by internal (eg, emotional state) or external (eg, sights, smells) cues (Figure 19.6). **Context-dependent effects** are external cues that aid retrieval. For example, if an individual encodes a memory at the library, that memory is easier to recall (ie, retrieve) at the library than in class. **State-dependent effects** are internal cues that aid retrieval. For example, if an individual encodes a memory while happy, that memory is easier to retrieve during a later happy mood than when sad.

State (mood) effects

Improved recall when the same internal cues are present at both encoding and retrieval

Context (environment) effects

Improved recall when the same external cues are present at both encoding and retrieval

Figure 19.6 State-dependent versus context-dependent memory effects.

The **serial position effect** describes how the relative ease (or difficulty) of remembering an item from a list is related to the item's position on the list. The items that are easiest to recall are those from the beginning (called the primacy effect) and end (called the recency effect) of the list, while the middle items are the hardest to recall. For example, when an individual views a list of items one at a time, the items studied first and last would be the easiest to remember, and those from the middle would be the hardest to remember (Figure 19.7).

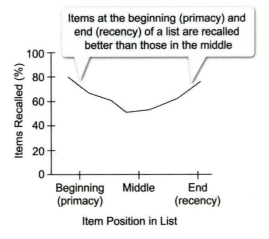

Figure 19.7 The serial position effect.

A retrieval failure in which a person experiences only partial recall of a word or term is called the **tip-of-the-tongue phenomenon**. Individuals experiencing the tip-of-the-tongue phenomenon can often recall details about the word they are trying to retrieve, such as the first letter or number of syllables, but they cannot recall the word itself despite feeling that the information is "on the tip of the tongue." For example, an individual is unable to recall the word "peony" despite knowing that the word is the name of a flower and begins with the letter "p" (Figure 19.8).

For example, while planning his flower garden, an individual cannot recall the name of a specific flower. Although he knows that the name begins with the letter "p," he is unable to produce the word "peony," demonstrating the tip-of-the-tongue phenomenon.

Figure 19.8 Tip-of-the-tongue phenomenon example.

Lesson 19.2
Forgetting

19.2.01 Decay and Interference

Memory Decay

Early memory researcher Hermann Ebbinghaus assessed his own memory by studying a list of short nonsense syllables and then repeatedly testing his memory for the syllables over time. Ebbinghaus found that **memory decay** (ie, forgetting) follows a characteristic pattern known as the **forgetting curve** (Figure 19.9): the initial rate of decay is greatest right after the material is first learned, then the rate of decay plateaus over time unless the material is reviewed again.

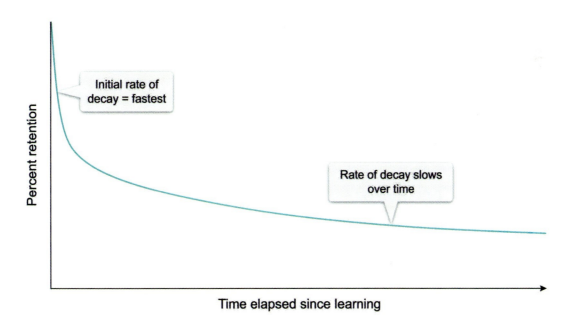

Figure 19.9 Typical memory decay: the forgetting curve.

Numerous studies have since produced the same basic forgetting curve shape for different types of memory, including short-term and several types of long-term memory (eg, episodic, semantic, procedural).

Interference

A common memory error that occurs when previously learned information interferes with the ability to recall new information is called **proactive interference**. For example, an individual's father cannot remember her new boyfriend's name (ie, recent information) and repeatedly refers to him by her old boyfriend's name (ie, older information).

Conversely, **retroactive interference** occurs when recently encoded information prevents the recall of older information. For example, an individual's father cannot remember her old boyfriend's name (ie, older information) and repeatedly refers to him by her new boyfriend's name (ie, recent information). Proactive and retroactive interference are contrasted in Figure 19.10.

Figure 19.10 Proactive versus retroactive interference.

19.2.02 Memory Construction

Memories are not perfect recordings of past events. **Memory reconstruction** refers to how memories are altered in the process of retrieval and subsequent storage.

In Elizabeth Loftus's memory reconstruction experiments, participants were shown films of car accidents and later asked about them. The results showed that a question about "the crash" as opposed to "the fender bender" influenced how the participants remembered the event. This **misinformation effect** explains how misleading information presented after an event can distort memories, and these experiments are often cited to bring the validity of eyewitness testimony into question.

The process of memory reconstruction leads to some common memory mistakes, such as **source monitoring errors**, which occur when a memory is attributed to the wrong source. For example, an individual may think they heard a joke from their father, but they actually heard it from a friend. The misinformation effect and source monitoring errors help explain **false memories**: memories that are distorted or memories of something that did not occur.

19.2.03 Aging and Memory

Aging affects the various types of memory differently. Aging has been associated with declines in certain types of memory, including episodic and source memory. Episodic memory (Concept 19.1.02) is the memory of autobiographical events (eg, the name of a childhood friend), whereas source memory is memory for the source of learned information (eg, correctly attributing one's knowledge of an event to a newspaper article).

Aging is also associated with declines in **flashbulb memory**: a vivid, detailed type of autobiographical memory for an event that was extremely emotional, distinct, or significant to the individual (eg, the 9/11 attacks, the birth of a child). Individuals are able to vividly recall specific details surrounding the event, such as what they were wearing and their emotional state at the time of the event. Once thought to be extremely accurate snapshots of emotionally arousing events, studies suggest that for people of all ages, flashbulb memories are not completely accurate or consistent over time, despite people's confidence in their recollections.

In contrast, other types of memory appear to remain relatively stable with age, such as semantic memory (memory for words, facts, and concepts that have been acquired over the lifetime) (see Concept 19.1.02). Additionally, procedural memory, which involves motor skills one has acquired, also remains relatively stable across adulthood.

19.2.04 Memory-Related Symptoms

Amnesia is severe memory loss that can be caused by brain trauma (eg, a head injury). Amnesia can be classified by whether the memory loss affects new or old explicit/declarative memories (ie, memory for facts and events that can be intentionally recalled):

- **Retrograde amnesia** is the loss of memories acquired prior to the trauma. For example, the patient cannot remember what they were doing just before the trauma.
- **Anterograde amnesia** is the inability to form new memories. For example, a patient can temporarily learn information while paying attention, but this information, such as a new plan, cannot be permanently stored and recalled later. Retrograde and anterograde amnesia are contrasted in Figure 19.11.

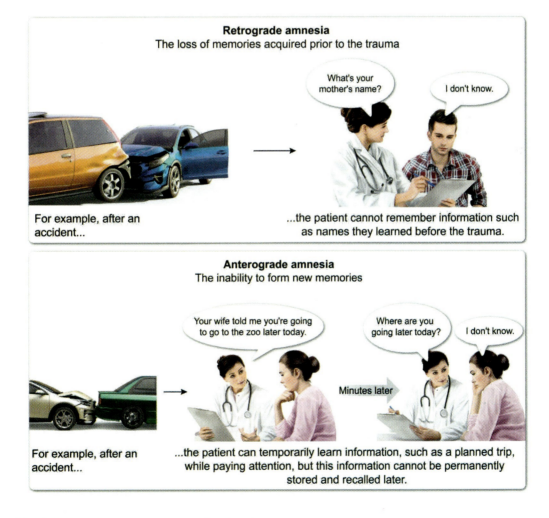

Figure 19.11 Retrograde versus anterograde amnesia.

Lesson 19.3
The Biological Underpinnings of Memory

19.3.01 Neural Plasticity

Neural plasticity (or neuroplasticity) refers to the ability of neurons to change. The strengthening of neural connections, known as potentiation, and weakening of neural connections, known as depression, illustrate plasticity. Neuroplasticity enables the modification of neurons during learning and can allow entire brain regions to recover function after an injury.

Plasticity allows for synaptic as well as structural changes. Synaptic plasticity is exemplified by changes in the firing rate of the presynaptic neuron(s) altering the amount of neurotransmitter released into the synaptic cleft and/or the number of postsynaptic receptors. In contrast, forms of structural plasticity include sprouting (new connections between neurons), rerouting (altered connections between neurons), and pruning (elimination of connections between neurons).

Mechanisms such as these can, in some cases, allow the brain to repurpose an area that is no longer used. For example, in an individual who lost their sight at an early age, the occipital lobe (visual processing area) is no longer receiving visual information. That brain area may then become involved in processing information from the other senses (eg, auditory processing) (Figure 19.12).

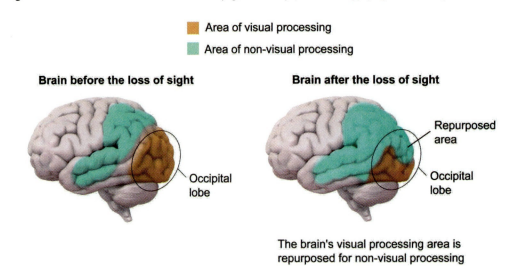

Figure 19.12 Example of neural plasticity after the loss of sight.

19.3.02 Long-Term Potentiation

As Concept 19.3.01 introduces, neural plasticity refers to neurons' ability to change (eg, alterations to synapses). **Long-term potentiation** (LTP) occurs when synapses that are stimulated frequently are strengthened (ie, become more effective). Animal research has supported the assertion that LTP enables both associative and non-associative learning.

Although LTP occurs at synapses throughout the brain, it has been extensively studied at glutamatergic synapses in the hippocampus because LTP at these synapses is hypothesized to be the mechanism underlying memory consolidation. Consolidation is the process of converting memories that are being kept temporarily as synaptic alterations into long-term memory. Researchers have shown that brief, high-frequency (ie, tetanic) stimulation of a hippocampal glutamatergic neuron induces LTP and an excitatory postsynaptic potential (see Concept 4.2.02) that is increased in magnitude relative to baseline (see Figure 19.13).

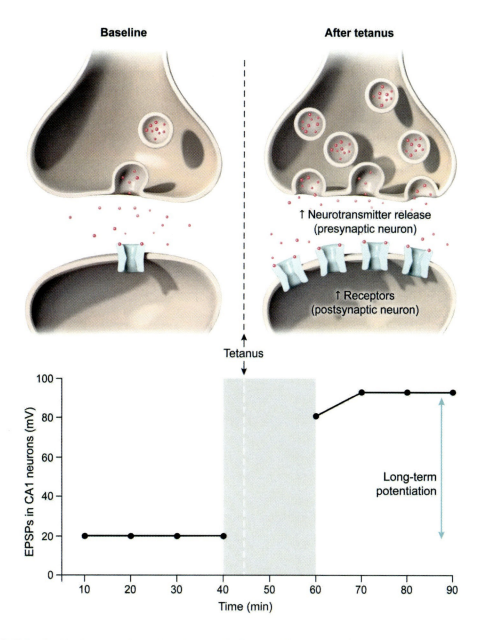

Figure 19.13 Tetanic stimulation induces long-term potentiation at hippocampal synapses.

LTP can result in changes to the presynaptic or postsynaptic neurons. Increased calcium influx causes increased presynaptic neurotransmitter release. At the postsynaptic neuron, LTP can cause phosphorylation of postsynaptic receptors (leading to increased effectiveness), the insertion of new receptors into the membrane, and an increase in dendritic spine density, for example. The exact mechanisms of LTP vary based on brain region and synapse type.

Increased presynaptic neurotransmitter release is a more immediate but transient form of LTP. Longer lasting strengthening of synaptic connections involves alterations in gene expression, the synthesis of new proteins, and/or new synaptic connections. The consolidation of short-term memory into long-term memory, for example, requires changes to existing proteins at synapses as well as new protein synthesis and altered gene expression.

In addition to occurring as a result of repeated, high-frequency stimulation from one presynaptic input, LTP can also occur when two (or more) presynaptic neurons repeatedly fire at the same time. For example, if neuron C repeatedly receives simultaneous input from two sources, neuron A and neuron B, both synapses may become potentiated. Following potentiation, either neuron A or neuron B alone can cause an action potential in neuron C, whereas previously neither was sufficient to induce postsynaptic depolarization to threshold. This process is hypothesized to be the neural foundation for learned associations, such as if, in this case, neuron A relays visual information about a flower and neuron B relays olfactory information about a flower, the appearance and smell of a flower become linked.

In contrast, **long-term depression** (LTD) describes when synapses that are stimulated infrequently are weakened. LTD can result in, for instance, a decrease in presynaptic neurotransmitter release, dephosphorylation and an internalization of postsynaptic receptors, a decrease in dendritic spine density, and synaptic pruning. Examples of LTP and LTD are depicted in Figure 19.14.

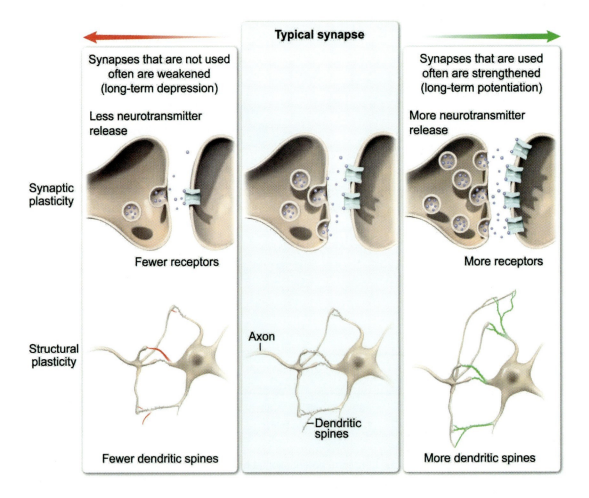

Figure 19.14 Examples of long-term potentiation and depression.

Lesson 20.1

Cognition Across the Lifespan

20.1.01 Piaget's Theory of Cognitive Development

Jean Piaget's theory of cognitive development states that humans progress through four distinct, age-related stages in which they master increasingly complex cognitive tasks.

During the **sensorimotor stage** (ages ~0–2), children explore the world using their senses (eg, touch) and motor movements (eg, grabbing). Attainment of **object permanence** occurs when the child becomes aware that something still exists even when it cannot be seen. For example, a child who looks under a bed for a ball she just saw roll underneath has developed object permanence (Figure 20.1).

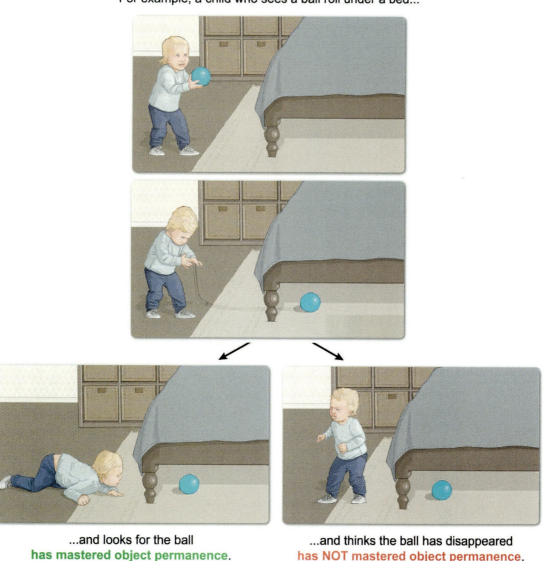

Figure 20.1 Object permanence example.

During the **preoperational stage** (ages ~2–7), children engage in pretend play and develop language. The preoperational stage is marked by **egocentrism**, the inability to assume another's point of view. For example, a child assumes that his favorite food is also his dad's favorite food.

During the **concrete operational stage** (ages ~7–11), children begin to think logically about concrete events and to master **conservation**: the understanding that an object's properties (eg, amount of water) remain the same even if its form changes (eg, water is poured from a tall glass to a wide glass) (Figure 20.2).

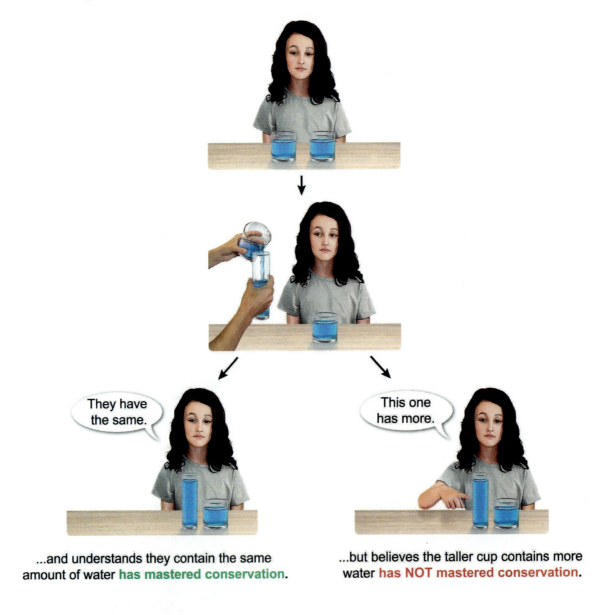

Figure 20.2 Conservation example.

During the **formal operational stage** (ages 11+), children develop moral reasoning and hypothetical thinking as a result of logical, abstract thinking. For example, a child in this stage uses hypothetical, abstract thinking to solve problems in algebra class (Figure 20.3).

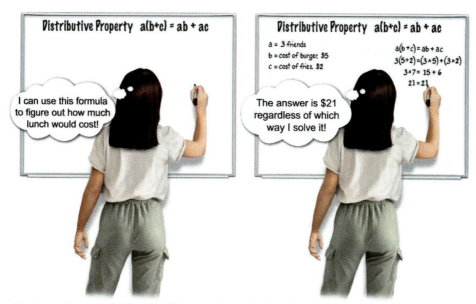

Figure 20.3 Formal operational stage example.

Piaget's Theory of Learning

Schemas are mental representations based on culture and experience that guide expectations. For example, a schema guides the expectations for student and teacher behavior in a classroom setting. According to Piaget, **assimilation** occurs when children interpret new information or experiences using existing schemas. For example, a child might call any animal in the water, even a mammal such as an otter, a "fish" (Figure 20.4).

Figure 20.4 Assimilation example.

In contrast, **accommodation** occurs when new information changes existing schemas. For example, a child learns that a pony is a short, adult horse (not a baby horse) and modifies her mental representation for horses to include short adult horses called ponies (Figure 20.5).

For example, when a child learns new information about ponies, her existing schema for horses is changed.

Figure 20.5 Accommodation example.

Lesson 20.2

Theories of Intelligence

20.2.01 Theories of Intelligence

Intelligence is the ability to learn, adapt, and solve problems. Some experts, such as Charles Spearman, asserted that intelligence reflects a single trait (ie, the **g factor**). This single trait is thought to underlie performance on the tasks found on standardized intelligence tests (eg, Wechsler, Stanford-Binet) and predict academic abilities.

In contrast, Howard Gardner's **theory of multiple intelligences** covers a broad range of skills in different domains. The different types of intelligence in Gardner's theory include linguistic, visual-spatial, musical-rhythmic, logical-mathematical, intrapersonal, interpersonal, naturalist, and kinesthetic (described in Figure 20.6).

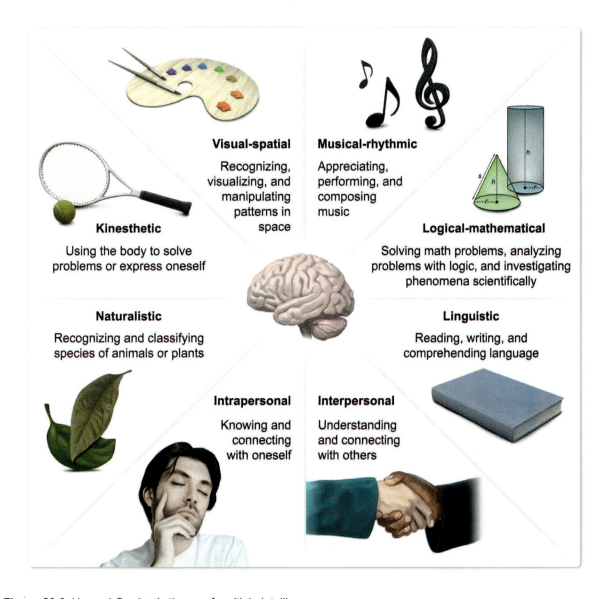

Figure 20.6 Howard Gardner's theory of multiple intelligences.

Another theory, Robert Sternberg's **triarchic theory of intelligence** (Figure 20.7), consists of three types of intelligence: practical (ie, applying real-world knowledge to manage everyday problems), creative (ie, managing novel situations and inventing new things), and analytical (ie, scrutinizing, evaluating, and solving problems).

Figure 20.7 Robert Sternberg's triarchic theory of intelligence.

Lesson 21.1
Types of Problem-Solving

21.1.01 Types of Problem-Solving

Problem-solving involves identifying a problem, coming up with a tactic or strategy to solve the problem, carrying out the tactic or strategy, and evaluating whether a solution has been attained.

There are several common problem-solving methods. **Trial and error** involves attempting possible solutions until the problem is solved, ruling out ineffective solutions along the way. For example, a psychiatrist may try various antidepressants for a patient, monitoring their effects over time until the best medication is found. Trial and error is most viable when there are a limited number of options.

The problem-solving strategy referred to as an **algorithm** is a systematic (ie, step-by-step) procedure (eg, using an algebraic formula) that produces an accurate solution to a well-defined problem (eg, an algebraic equation). Although a solution is guaranteed, algorithms may also be complex and time-consuming. See Figure 21.1 for an example of an algorithm.

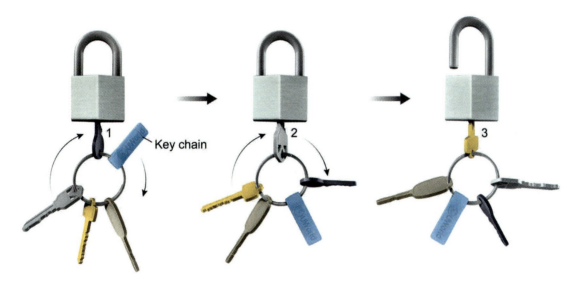

Unlocking a padlock by starting with the key closest to the key chain and systematically trying the next key and then the next one until finding the key that unlocks the padlock is an example of an algorithm.

Figure 21.1 Algorithm example.

In contrast, a **heuristic** is a mental shortcut that allows for fast problem-solving and decision-making. Heuristics are less time-consuming than algorithms. However, unlike algorithms, heuristics sometimes lead to inaccurate conclusions. For example, to calculate a 20% tip on a dinner bill, an individual could apply a heuristic and simply double the 8% sales tax listed on the bill (instead of using an algorithm such as a mathematical formula to produce the correct answer). This heuristic would be faster but less accurate than the algorithm (ie, 16% as compared to 20%). Heuristics are described in more detail in Concept 21.2.01.

Lesson 21.2

Barriers to Effective Problem-Solving

21.2.01 Heuristics

As Concept 21.1.01 introduces, heuristics are mental shortcuts that allow for fast problem-solving and decision-making but sometimes lead to inaccurate conclusions. Examples of heuristics include the representativeness heuristic and the availability heuristic.

The **representativeness heuristic** is the tendency to compare things (eg, people, events) to mental prototypes (ie, typical or standard examples) when making judgments. For example, a patient assumes that a male clinician is a doctor (and not a nurse) because the clinician matches the patient's mental prototype that male clinicians are doctors.

The **availability heuristic** is the tendency to believe that if something is easily recalled from memory, it must be common or likely. For example, an individual easily recalls news coverage of shark attacks and then incorrectly assumes shark attacks are common (Figure 21.2).

Figure 21.2 Availability heuristic example.

21.2.02 Biases

Cognitive biases are common problem-solving obstacles that result in illogical conclusions. **Confirmation bias** occurs when an individual has a preconceived belief and looks only for evidence supporting that belief, ignoring contradictory evidence. For example, an individual only consumes media (eg, articles, TV) that support their political views, subsequently ignoring contrary information (Figure 21.3).

Chapter 21: Problem-Solving and Decision-Making

One might consume media (eg, social media) that support only one's political views, subsequently ignoring contrary information.

Figure 21.3 Example of confirmation bias.

In contrast, **hindsight bias** occurs when an event is perceived as having been predictable after it occurs. For example, an individual declares that they knew "all along" that they were going to run out of gas (see Figure 21.4).

For example, after running out of gas, an individual declares that she knew "all along" that they were going to run out of gas and should have gotten gas at the last station.

Figure 21.4 Example of hindsight bias.

21.2.03 Other Barriers

Other common barriers to effective problem-solving include mental set and functional fixedness.

Mental set describes when an individual continues using a problem-solving method that worked previously but is not right for the current problem. For example, after opening several doors by pulling the handle, an individual may repeatedly pull a door handle that must be pushed to open.

Similarly, **functional fixedness** prevents an individual from thinking of different uses for an object. For example, if an individual who needs to tighten a screw does not have a screwdriver and instead uses the edge of a coin, that individual has overcome functional fixedness (Figure 21.5).

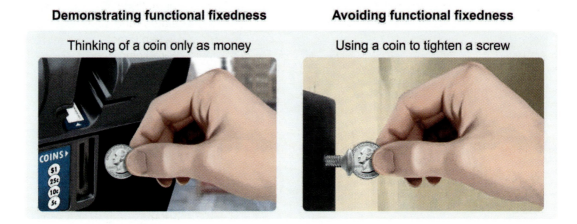

Figure 21.5 Example of functional fixedness.

Lesson 22.1
Theories of Language Development

22.1.01 Learning Theory

There are several language development theories; they differ in the extent to which language acquisition is characterized as learned (eg, learning perspective) versus innate (eg, nativist perspective).

B.F. Skinner's **learning theory** (also known as the behaviorist theory) argues that language is an entirely learned behavior. This theory suggests that humans are born as "blank slates" and develop language skills through operant conditioning, imitation, and practice. For example, when infants make vocalizations that sound like "mama," they are rewarded with attention and affection, thereby increasing the likelihood that that behavior will be repeated. However, when infants make other sounds that are not similar to words, they do not receive reinforcement.

22.1.02 Nativist Theory

In contrast to the learning theory (Concept 22.1.01), the **nativist theory** of language development proposes that language is not learned like other behaviors are (ie, through conditioning and modeling); instead, the learning of language is an innate process hardwired in the brain.

A proponent of the nativist theory, linguist and cognitive scientist Noam Chomsky argued for the existence of a hypothetical **language acquisition device** (LAD) that innately prewires the human brain to learn language. He stated that it is the existence of the LAD that allows children to readily pick up language from their caregivers and other people.

Research supports the existence of a critical period of language development (Figure 22.1), which suggests that there is a time-sensitive period early in life during which language acquisition is easier (with proper exposure), as compared to the period afterward, during which language acquisition is much more difficult. The nativist theory asserts that humans will learn whichever language(s) they are exposed to during the critical period. Support for this theory includes that certain brain regions involved in language development (eg, Wernicke's area) are similar in all humans.

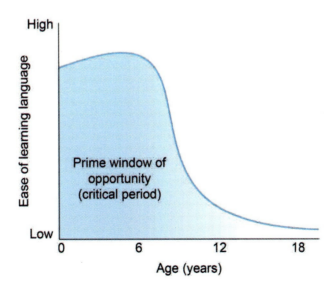

Figure 22.1 Critical period for language development.

22.1.03 Interactionist Theory

The **interactionist theory** proposes that language acquisition is the result of both biological factors (eg, typical brain development) and environmental factors (eg, the interaction that occurs between children and their caregivers).

The interactionist theory is supported by evidence that certain aspects of language appear to be innate whereas others appear to be social. Children typically learn to communicate with language along a similar timeline (eg, first words around age one, simple two-word phrases by age two), which provides evidence that some aspects of language acquisition are innate. However, children who are severely neglected (ie, almost no social contact) do not learn to communicate using language, which provides evidence that language acquisition also requires social interaction.

The learning (Concept 22.1.01), nativist (Concept 22.1.02), and interactionist theories are compared in Table 22.1.

Table 22.1 Theories of language development.

Theory	Language acquisition

Learning perspective	• Learned via:
	  ○ Operant conditioning
	  ○ Language imitation and practice

Nativist perspective	• Innate and biologically predetermined ○ Occurs before a critical (time-sensitive) period in early life
Interactionist perspective	• Biological (due to normal brain development) **AND** • Social (due to interaction, reinforcement, and motivation to communicate)

Lesson 22.2

Language and Cognition

22.2.01 Language and Cognition

The Sapir-Whorf hypothesis, also known as the **linguistic relativity hypothesis**, argues that language influences perception and cognition. For example, if an individual does not have vocabulary for a "toe-loop" ice-skating jump or a "lutz" jump (and instead refers to each as a "jump"), that person will find it difficult to tell the two jumps apart (Figure 22.2).

For example, the linguistic relativity hypothesis predicts that if a person does not have vocabulary for a "toe loop" ice skating jump or a "lutz" jump (and refers to them both as "jump"), that person will find it difficult to tell them apart.

Figure 22.2 Linguistic relativity hypothesis example.

Linguistic determinism, a stronger version of this hypothesis, states that language determines perception and cognition. For example, if an individual does not have vocabulary for a "toe-loop" ice-skating jump or a "lutz" jump (and instead refers to each as a "jump"), linguistic determinism predicts that this person would be unable to tell the two jumps apart.

Lesson 22.3

The Biological Underpinnings of Language and Speech

22.3.01 The Biological Underpinnings of Language and Speech

Two key language centers in the brain, Broca's area and Wernicke's area, are named after the researchers who identified the functions of these specialized regions (Figure 22.3). Paul Broca discovered that a region now called **Broca's area**, located in the left frontal lobe in most people, is responsible for language production. Damage to Broca's area results in a type of aphasia (a problem with language production or comprehension) in which patients have difficulty producing spoken or written language (eg, mispronouncing words).

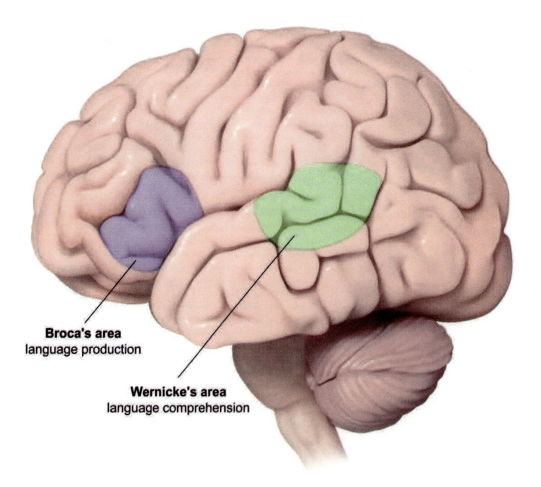

Figure 22.3 Broca's and Wernicke's areas.

Carl Wernicke discovered that a region now called **Wernicke's area**, located in the left temporal lobe in most people, is responsible for language comprehension. Damage to Wernicke's area results in a type of aphasia in which patients have difficulty comprehending spoken and written language (eg, difficulty understanding what others are saying).

END-OF-UNIT MCAT PRACTICE

Congratulations on completing **Unit 4: Learning, Memory, and Cognition**.

Now you are ready to dive into MCAT-level practice tests. At UWorld, we believe students will be fully prepared to ace the MCAT when they practice with high-quality questions in a realistic testing environment.

The UWorld Qbank will test you on questions that are fully representative of the AAMC MCAT syllabus. In addition, our MCAT-like questions are accompanied by in-depth explanations with exceptional visual aids that will help you better retain difficult MCAT concepts.

TO START YOUR MCAT PRACTICE, PROCEED AS FOLLOWS:

1) Sign up to purchase the UWorld MCAT Qbank
   IMPORTANT: You already have access if you purchased a bundled subscription.
2) Log in to your UWorld MCAT account
3) Access the MCAT Qbank section
4) Select this unit in the Qbank
5) Create a custom practice test

Unit 5 Motivation, Emotion, Attitudes, Personality, and Stress

Chapter 23 Motivation

23.1 Influences on Motivation

23.1.01	Instinct	
23.1.02	Drives	
23.1.03	Arousal Theory	
23.1.04	Needs	

23.2 Theories of Motivation

23.2.01	Drive Reduction Theory	
23.2.02	Incentive Theory	
23.2.03	Other Theories	

Chapter 24 Emotion

24.1 The Principles and Components of Emotion

24.1.01	Components of Emotion	
24.1.02	Universal Emotions	
24.1.03	Adaptive Function of Emotions	

24.2 Theories of Emotion

24.2.01	Theories of Emotion	

24.3 The Biological Underpinnings of Emotion

24.3.01	The Biological Underpinnings of Emotion	

Chapter 25 Attitudes

25.1 Attitudes and Behavior

25.1.01	Influence of Behavior on Attitudes	
25.1.02	Cognitive Dissonance Theory	

Chapter 26 Personality Theories

26.1 Psychoanalytic Theories of Personality

26.1.01	Psychoanalytic Theories of Personality	

26.2 Humanistic Theories of Personality

26.2.01	Humanistic Theories of Personality	

26.3 Trait Theories of Personality

26.3.01	Trait Theories of Personality	

Chapter 27 Stress

27.1 The Principles and Components of Stress

27.1.01	Appraisal Theory	
27.1.02	Types of Stressors	

27.2 The Effects of Stress and Stress Management

    27.2.01   Effects of Stress
    27.2.02   Managing Stress

Chapter 28 Theories of Attitude and Behavior Change

28.1 Theories of Attitude and Behavior Change

    28.1.01   Elaboration Likelihood Model
    28.1.02   Social Cognitive Theory

Lesson 23.1
Influences on Motivation

23.1.01 Instinct

Motivation describes the factors that prompt action toward a goal. Evolutionary theorists claim that behaviors are motivated by instincts. An **instinct** is an innate, fixed pattern of behavior that is more complex than a reflex, which is a simple response to a stimulus (eg, jerking one's hand away from a hot stove). Instincts are not based on prior experience or learning. For example, newly hatched sea turtles instinctively know to move toward the ocean and swim.

23.1.02 Drives

A **drive** is an internal state that motivates an organism to fulfill a need. Biological needs (eg, food) resulting from physiological changes (eg, a decrease in glucose levels) create drives. Drives (eg, hunger) then prompt action (eg, food-seeking behaviors). Drives are discussed further in Concept 23.2.01.

23.1.03 Arousal Theory

The **arousal theory** of motivation suggests that individuals are motivated to maintain an optimum level of arousal. For example, if bored at home, one will go out dancing, or if overwhelmed at a party, one will step outside for a quieter setting.

Researchers have identified a relationship between arousal and performance. The **Yerkes-Dodson law** indicates that there is an optimal level of physiological or mental arousal at which performance is maximized; performance will decline with too little or too much arousal. For example, athletes tend to perform best when nervous but not too nervous.

23.1.04 Needs

Some psychologists classify factors that prompt behaviors (ie, motivation) as types of needs. Researchers have identified several psychological needs, including:

- The need for power: a desire to influence other people and have control over them.
- The need for affiliation: a desire for positive social relationships and interactions.
- The need for achievement (also known as achievement motivation): a desire to improve upon one's past performance, compete with high standards, and attain significant accomplishments.

Lesson 23.2
Theories of Motivation

23.2.01 Drive Reduction Theory

As Concept 23.1.02 introduces, a drive is an internal state that prompts action. The **drive reduction theory** proposes that motivation results from a disruption of homeostasis. Homeostasis refers to physiological equilibrium (balance). Certain physiological changes (eg, a decrease in glucose levels) disrupt homeostasis, creating a biological need. A drive (eg, hunger) prompts the organism to fulfill that need (eg, to eat) and restore homeostasis (eg, normal glucose levels). An example of drive reduction is depicted in Figure 23.1.

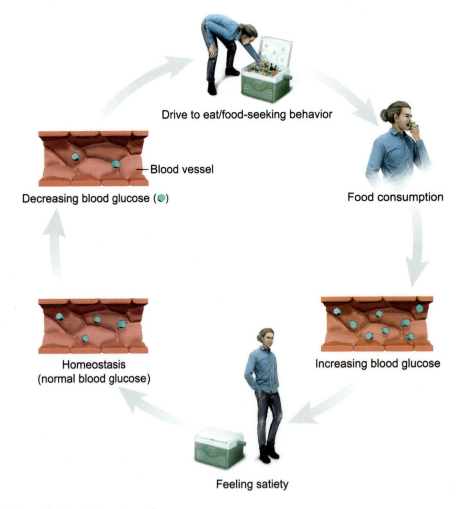

Figure 23.1 Drive reduction theory example.

23.2.02 Incentive Theory

Behavior that is primarily motivated by external rewards (eg, money), external pressures (eg, deadlines), or biological drives (eg, hunger) can be classified as **extrinsically motivated**. In contrast, **intrinsic motivation** refers to behavior that is primarily motivated by internal rewards (eg, enjoyment). Examples of intrinsic and extrinsic motivation are illustrated in Figure 23.2.

Extrinsic motivation: when external rewards or pressures (eg, grades, money), or biological drives (eg, hunger) motivate behavior

Intrinsic motivation: when internal rewards (eg, enjoyment) motivate behavior

For example, an individual studies (ie, behavior) because she...

...has an exam in two days (ie, external pressure).

...enjoys the subject (ie, internal reward).

Figure 23.2 Extrinsic versus intrinsic motivation.

The **incentive theory** of motivation suggests that organisms are motivated to act to obtain external rewards (ie, incentives) or to avoid negative consequences (eg, punishment). For example, an individual might complete chores, not because they enjoy cleaning (an internal factor), but because they want to get money from their grandfather for completing the chores (an external reward).

23.2.03 Other Theories

Expectancy Theory

The **expectancy theory** of motivation proposes that individuals are motivated to act based on the expected outcomes of their behavior. According to this theory, motivation involves expectancy, instrumentality, and valence:

- Expectancy is the belief that one will be able to achieve the desired outcome. Asking students to rate how successful they think they will be on an exam is a measure of expectancy.
- Instrumentality is the belief that one has control over the desired outcome. Asking students to rate how much control they believe they have over their success on an exam is a measure of instrumentality.
- Valence involves the value placed on the desired outcome. Asking students to rate how much they want to succeed on an exam is a measure of valence.

The Hierarchy of Needs

Lastly, as Lesson 26.2 covers, Abraham Maslow's **hierarchy of needs** proposes that humans are motivated to achieve needs in a hierarchy of importance; lower needs (eg, shelter) must be met before higher needs (eg, esteem).

Lesson 24.1

The Principles and Components of Emotion

24.1.01 Components of Emotion

The three components of emotion are:

- The **cognitive** component, which includes all the mental processes that accompany the emotion (eg, thoughts, evaluation of the context).
- The **behavioral** component, which is the immediate outward expression that occurs in response to an emotion. These responses are typically involuntary (eg, smiling, gasping).
- The **physiological** component, which includes all the bodily processes that accompany the emotion (eg, changes in heart rate, sweating).

24.1.02 Universal Emotions

The **universal emotion theory** states that certain emotions are expressed and detected by everyone, regardless of culture. According to this theory, the universal emotions include happiness, sadness, fear, anger, disgust, and surprise (Figure 24.1).

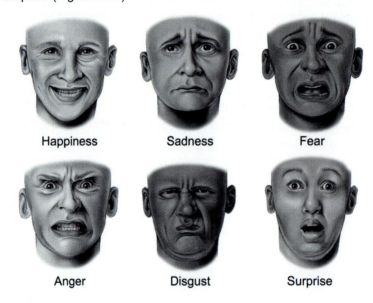

Figure 24.1 Universal emotions.

In Paul Ekman's experiments, individuals from different cultures were shown photographs of people displaying different facial expressions. Regardless of their cultural background, participants were able to identify universal emotions from the photographs.

24.1.03 Adaptive Function of Emotions

Emotions and empathy play an adaptive role by allowing humans to emotionally connect with others and by fostering group cohesion. A desire to experience certain emotions (eg, happiness, pride) and avoid other emotions (eg, embarrassment, shame) causes people to behave in predictable, socially acceptable ways. Understanding the emotions of others (ie, empathy) and expressing emotions to others are important elements of social interaction.

Lesson 24.2
Theories of Emotion

24.2.01 Theories of Emotion

Theories of emotion attempt to explain why emotions exist and how they are generated and experienced. There are several major theories of emotion, including the James-Lange, Cannon-Bard, and Schachter-Singer theories.

The **James-Lange theory** of emotion states that specific physiological responses (eg, racing heart, sweating) produce specific emotions (eg, fear). For example, when approached by a snarling dog, a person will experience a racing heart (ie, physiological arousal), and this causes the feeling of fear (ie, subjective emotion).

In contrast, the **Cannon-Bard theory** of emotion states that physiological arousal and subjective emotion occur simultaneously. For example, when approached by a snarling dog, a person will simultaneously experience a racing heart (ie, physiological arousal) and the feeling of fear (ie, subjective emotion).

Finally, the **Schachter-Singer theory** (also called the two-factor theory) of emotion states that physiological arousal followed by cognitive appraisal (ie, interpretation) of that arousal (eg, "this situation is dangerous; my pounding heart signifies fear") produces emotions. For example, when approached by a snarling dog, a person will experience a racing heart (ie, physiological arousal) and interpret or label that arousal through cognitive appraisal, which causes the feeling of fear (ie, subjective emotion).

The Schachter-Singer theory addresses a limitation of the James-Lange theory: that most emotions correspond with nearly identical physiological responses. In other words, anger, fear, and excitement might all produce similar elevations in heart rate, but the experience of each emotion is quite different because emotion is the result of two factors, physiological arousal and cognitive interpretation of the situation. The James-Lange, Cannon-Bard, and Schachter-Singer theories of emotion are contrasted in Figure 24.2.

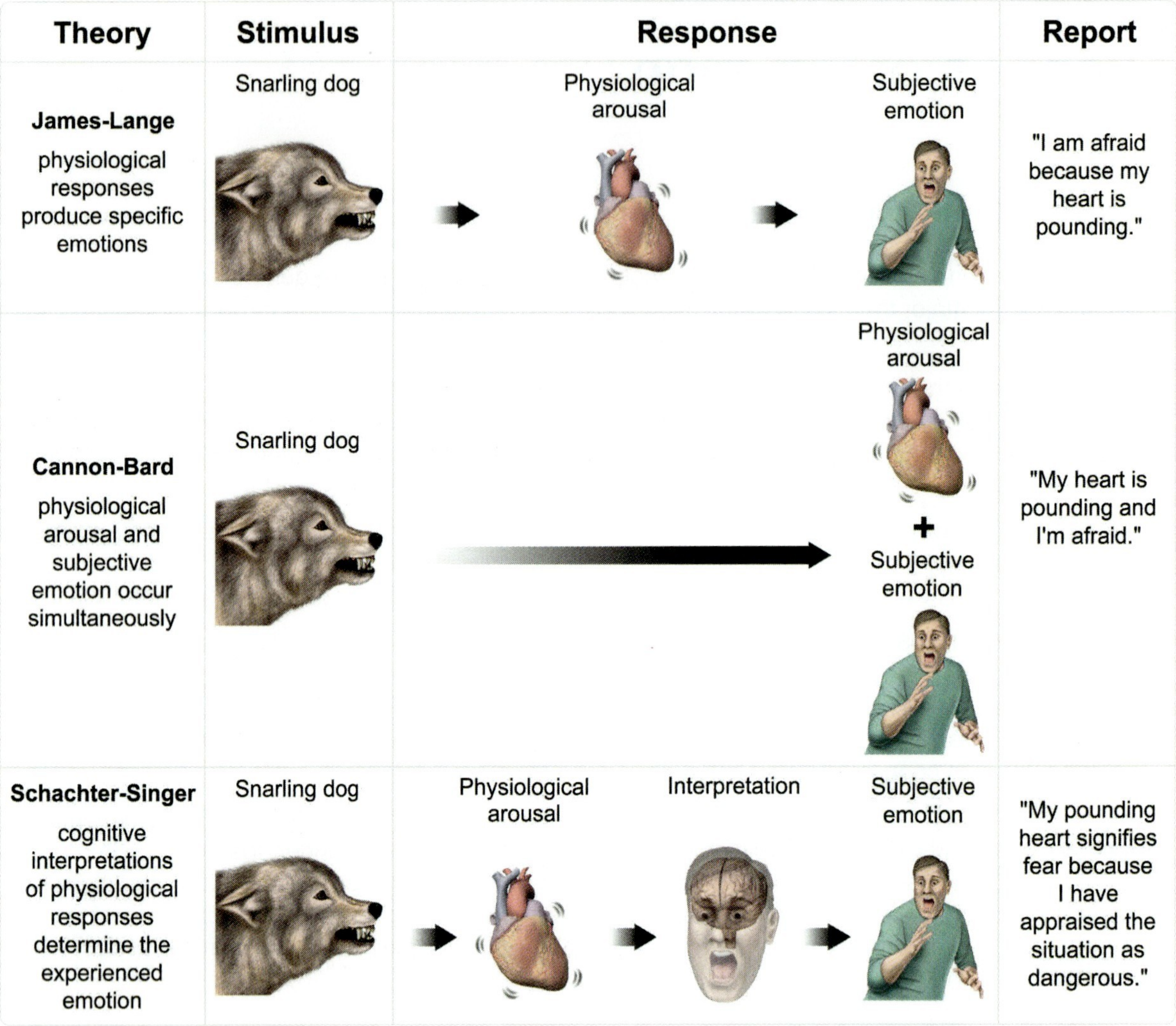

Theory	Stimulus	Response			Report
James-Lange physiological responses produce specific emotions	Snarling dog	Physiological arousal	→	Subjective emotion	"I am afraid because my heart is pounding."
Cannon-Bard physiological arousal and subjective emotion occur simultaneously	Snarling dog			Physiological arousal + Subjective emotion	"My heart is pounding and I'm afraid."
Schachter-Singer cognitive interpretations of physiological responses determine the experienced emotion	Snarling dog	Physiological arousal	Interpretation	Subjective emotion	"My pounding heart signifies fear because I have appraised the situation as dangerous."

Figure 24.2 Contrasting the James-Lange, Cannon-Bard, and Schachter-Singer theories of emotion.

Lesson 24.3

The Biological Underpinnings of Emotion

24.3.01 The Biological Underpinnings of Emotion

As Concept 4.3.01 introduces, several forebrain structures contribute to emotion, memory, and motivation; these brain areas are sometimes collectively referred to as the limbic system. Limbic structures particularly involved in emotion include:

- The **amygdala**, which plays a role in aggression and emotions such as fear (Figure 24.3). The amygdala is involved in fear conditioning and the formation of other emotionally charged memories. Researchers have shown that electrical stimulation of the amygdala can lead to displays of fear and aggression, whereas damage to the amygdala can result in a lack of fear.

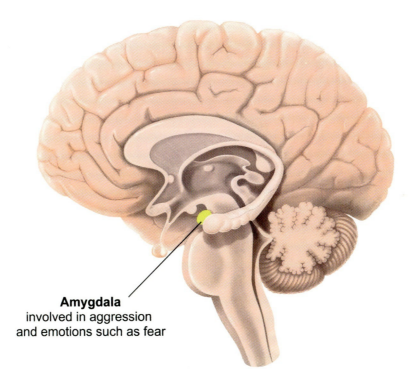

Figure 24.3 The amygdala.

- The **hypothalamus**, which releases hormones and controls the pituitary gland's hormone release (Figure 24.4); it coordinates many bodily processes such as hunger, growth, and the fight-or-flight stress response, as Lesson 4.5 discusses. The hypothalamus contributes to the physiological component of emotion (eg, changes in heart or respiration rate) by acting on the pituitary gland and the autonomic nervous system.

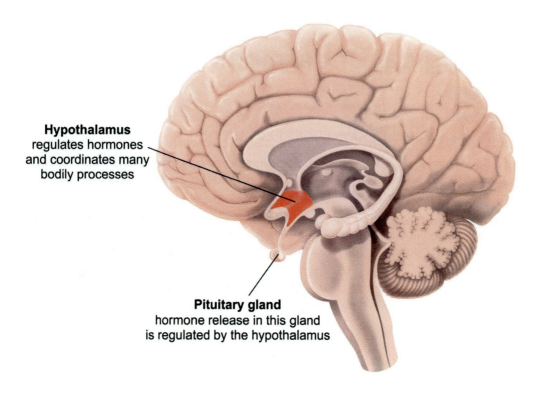

Figure 24.4 The hypothalamus.

Lesson 25.1

Attitudes and Behavior

25.1.01 Influence of Behavior on Attitudes

An **attitude** refers to an individual's evaluation of or inclination towards something (eg, another person, an object, an event). Attitudes can be positive, negative, or neutral and can change over time. Psychologists have identified three components of attitudes:

- The affective component involves how a person feels about something, including positive or negative emotions (eg, excitement, anger). For example, a person might have strong negative feelings about a political candidate, causing them to favor that candidate's opponent.

- The cognitive component involves a person's beliefs and opinions about something. For example, a person might believe that a particular political candidate has characteristics or experiences that make that candidate well-suited to hold office.

- The behavioral component involves how a person acts toward something. In the above examples, voting for or against a political candidate reflects the behavioral component of attitude.

A **role** (also called a social role) refers to the specific expected behaviors that correspond to a particular status in society (see Concept 36.3.01). In studies that ask participants to assume certain social roles, the behaviors that align with those roles have been shown to impact the participants' attitudes.

In the Stanford prison experiment (Figure 25.1), researchers assigned participants to the role of prisoner or guard and observed their behavior over several days in a prison-like environment. The experiment revealed that the participants changed their behavior to conform to the assigned roles and that the participants' behavior began to influence their attitudes.

For example, over the course of the study, the participants in the role of guard became increasingly hostile toward their peers assigned the role of prisoner. This hostility occurred to such an extent that the study was terminated prematurely out of concern for the participants' safety.

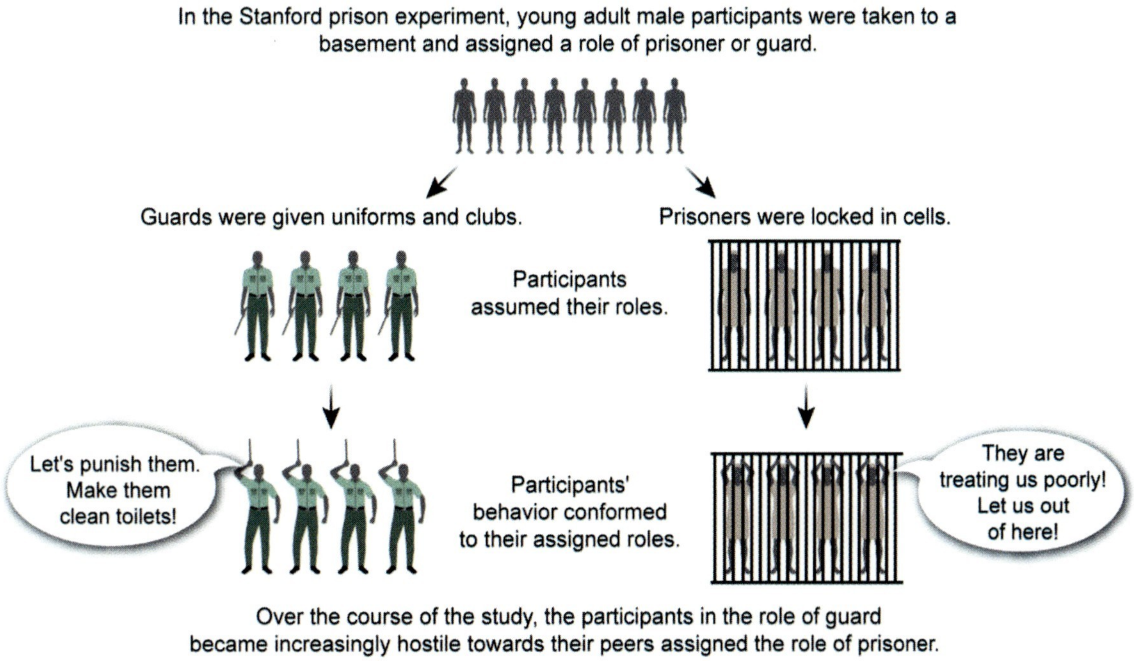

In the Stanford prison experiment, young adult male participants were taken to a basement and assigned a role of prisoner or guard.

Guards were given uniforms and clubs.   Prisoners were locked in cells.

Participants assumed their roles.

Let's punish them. Make them clean toilets!

Participants' behavior conformed to their assigned roles.

They are treating us poorly! Let us out of here!

Over the course of the study, the participants in the role of guard became increasingly hostile towards their peers assigned the role of prisoner.

Figure 25.1 The Stanford prison experiment.

Convincing people to change their attitudes or behaviors is known as persuasion. Two examples of persuasion techniques in which behavior influences attitudes are (Figure 25.2):

- The **foot-in-the-door strategy**, which involves first posing a small or easy request (eg, permission to go on a trip) and then when it is granted, posing a much bigger request (eg, chaperoning the trip). Individuals are more likely to agree to the larger request after agreeing to a smaller one.

- The **door-in-the-face strategy**, which involves first posing a big request (eg, chaperoning a trip) and then when it is declined, posing a much smaller or easier request (eg, permission to go on the trip). Individuals are more likely to agree to the smaller request after declining a larger one.

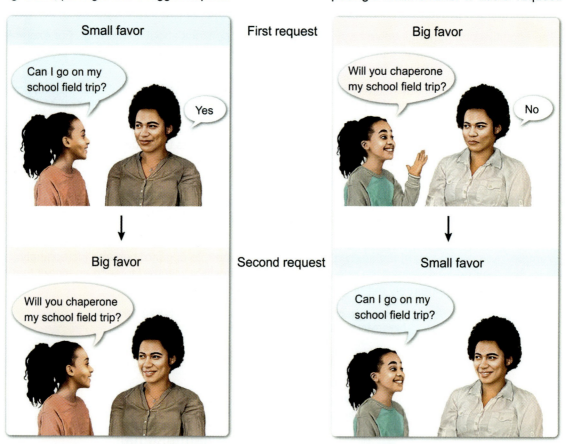

Figure 25.2 The foot-in-the-door versus the door-in-the-face strategies.

25.1.02 Cognitive Dissonance Theory

Cognitive dissonance occurs when contradictory thoughts and/or behaviors cause mental discomfort, which results in motivation to reduce the discomfort by aligning those thoughts and/or behaviors. For example, if a health-conscious individual knows there are negative health consequences of eating fast food but wants to continue eating burgers, he may experience psychological tension. To reduce this mental discomfort, the individual may try to convince himself that burgers are not that unhealthy (Figure 25.3).

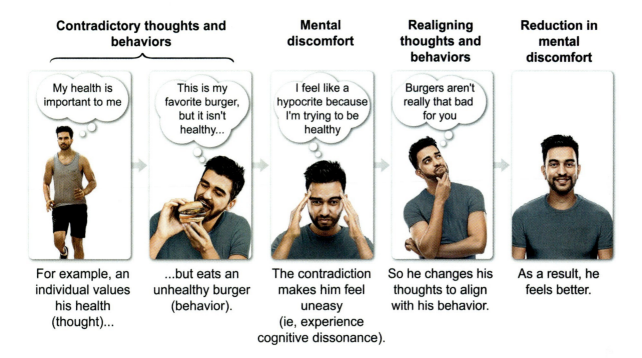

Figure 25.3 Cognitive dissonance example.

Lesson 26.1

Psychoanalytic Theories of Personality

26.1.01 Psychoanalytic Theories of Personality

Sigmund Freud

As Lesson 1.1 introduces, Sigmund Freud's **psychoanalytic theory** focuses on the impact of **unconscious** factors (eg, drives, conflicts stemming from childhood) on human development and behavior. Freud asserted that the three personality structures—id, ego, and superego—exist at different levels of conscious awareness (Figure 26.1).

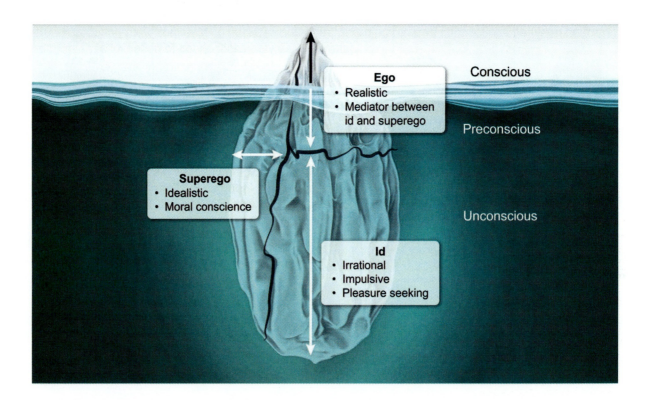

Figure 26.1 Sigmund Freud's model of personality and consciousness.

The **id** is impulsive, irrational, and pleasure-seeking, the **ego** is realistic and acts as a mediator between the id and superego, and the **superego** is the idealistic moral conscience. Whereas the id is entirely unconscious (beyond conscious awareness), the ego and superego span the conscious, preconscious (just beneath conscious awareness), and unconscious mind.

Freud asserted that personality results from conflict between inner desires and social restraints (eg, one's sense of right and wrong conflicting with one's impulses). One way the ego copes with this tension and protects itself from anxiety is through **defense mechanisms.** Defense mechanisms distort reality unconsciously and automatically. See Table 26.1 for a summary of selected common defense mechanisms.

Table 26.1 Major defense mechanisms (incomplete list).

Defense mechanism	Definition and example
Denial	Refusing or being unable to recognize unacceptable thoughts/behaviors (eg, insisting one is not angry when actually angry)
Projection	Attributing unacceptable thoughts/behaviors to someone or something else (eg, calling the sidewalk "stupid" after tripping)
Rationalization	Making excuses for unacceptable thoughts/behaviors (eg, justifying cheating because "the course is impossible")
Regression	Behaving as if much younger to avoid unacceptable thoughts/behaviors (eg, moving back in with parents to avoid personal responsibilities)
Repression	Blocking unacceptable thoughts/behaviors from consciousness (eg, being unaware of a traumatic past experience)
Displacement	Taking out unacceptable thoughts/behaviors on a safe target (eg, punching a pillow when angry at parents)
Sublimation	Transforming unacceptable thoughts/behaviors into acceptable thoughts/behaviors (eg, taking up boxing as a way to channel one's anger)
Reaction formation	Behaving in a manner opposite unacceptable thoughts/behaviors (eg, expressing love for a person one despises)

Carl Jung

Carl Jung was a neo-Freudian (early psychoanalyst) who stated that the mind contains both a personal unconscious (as described by Freud) and a shared, collective unconscious. The **collective unconscious** contains a store of inherited images, called **archetypes**, derived from our ancestors (eg, the figure of a wise old man is found in the folklore of many cultures).

In Jung's theory, the **self** is seen as the place where the personal unconscious, the conscious mind, and the collective unconscious meet. The archetypes found in the collective unconscious apply to one's personality. For example, the shadow, which contains the undesirable and shameful aspects of oneself, resides in the unconscious.

Alfred Adler

Another neo-Freudian, Alfred Adler, emphasized the impact of children's early **feelings of inferiority** on their personality. Adler stated that children feel inferior to adults (eg, adults are physically larger), as well as feel inferior to their siblings according to their birth order. For example, Adler would say that oldest children often overachieve (for instance, in school) because they are overcompensating for feeling inferior to younger children, who often receive more attention from their parents (Figure 26.2).

Chapter 26: Personality Theories

For example, an oldest child notices that his parents haven't paid as much attention to him now that he has a newborn brother and he feels unimportant.

As a result, the child tries to earn his parents' attention by doing exceptionally well in school.

Figure 26.2 Selected aspects of Alfred Adler's theory of personality.

Lesson 26.2

Humanistic Theories of Personality

26.2.01 Humanistic Theories of Personality

As Lesson 1.3 introduces, the **humanistic perspective** is based on the belief that humans are driven to achieve higher pursuits, such as **self-actualization** (ie, fulfilling one's greatest potential) and personal growth. Humanistic psychologists Abraham Maslow and Carl Rogers described different factors that they believed influence an individual's ability to attain self-actualization.

Abraham Maslow

Humanistic psychologist Abraham Maslow proposed that humans are motivated to achieve needs in a hierarchy of importance (Figure 26.3). These needs are:

- Basic physiological needs, which include satisfying hunger, thirst, and fatigue (eg, a runner stops to drink water during a marathon)
- Safety needs, which describe the need to feel secure and out of danger (eg, a young adult moves from a high-crime city center to a low-crime suburb)
- Belongingness needs, which describe the need for love and community (eg, a teenager joins a social club to feel like they fit in with their peers)
- Esteem needs, which describe the need for achievement and the need to be valued (eg, pursuing higher education for a sense of accomplishment)
- Self-actualization needs, which describe the need for fulfillment and the need to realize one's potential (eg, a successful lawyer with rewarding relationships fulfills their full potential by creating lasting, positive change in their community)

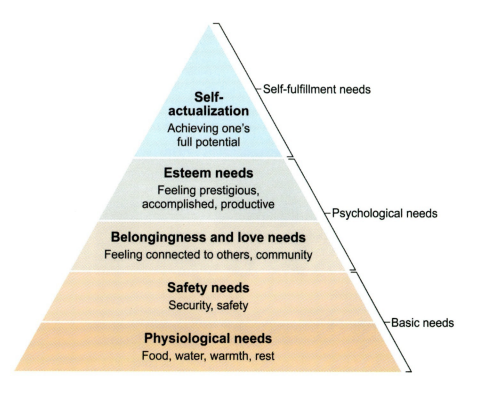

Figure 26.3 Abraham Maslow's hierarchy of needs.

According to Maslow, basic physiological needs must be met before psychological needs can be addressed, and in turn, psychological needs must be met before self-fulfillment needs can be addressed. Although Maslow believed that every person is capable of attaining self-actualization, he held that progress toward self-actualization would be thwarted by failure to attain lower needs.

Carl Rogers

Fellow humanistic psychologist Carl Rogers posited that **self-concept**, a person's ideas and feelings about who they are, is a primary part of personality. People whose *ideal self* (ie, the idea of who they should be) matches with their *actual* experiences will have a more positive self-concept. Positive self-concepts are linked to better functioning and health.

Rogers also believed that individuals' personal growth is influenced by their interactions with other people. Rogers described how **unconditional positive regard**, acceptance/support regardless of behavior, facilitates personal growth and progress toward self-actualization. Alternatively, receiving **conditional positive regard**, acceptance/support based on what a person does (eg, getting good grades), inhibits one's personal growth and progress toward self-actualization. Figure 26.4 depicts the interaction of some aspects of Carl Rogers's theory of personality.

Figure 26.4 Selected aspects of Carl Rogers's theory of personality.

Lesson 26.3
Trait Theories of Personality

26.3.01 Trait Theories of Personality

Trait theory describes personality as formed by enduring personal characteristics, or traits. **Traits**, which are stable over time, are the characteristic ways an individual thinks, feels, and acts. For example, an individual who is "friendly" will likely continue to act warm and affectionate (ie, friendly) to others (Figure 26.5).

For example, an individual who is "outgoing" in young adulthood will likely be "outgoing" later in life.

An individual who is "introverted" in young adulthood will likely be "introverted" later in life.

Figure 26.5 Examples of personality traits.

Paul Costa and Robert McCrae's **Big Five theory** is a trait theory that identifies five dimensions of personality: openness to experience (eg, curious, imaginative), extraversion (eg, outgoing, energetic), conscientiousness (eg, organized, responsible), agreeableness (eg, friendly, cooperative), and neuroticism (eg, moody, emotionally unstable). See Table 26.2 for a summary of these traits.

Table 26.2 Big Five personality traits.

Big Five trait	Examples of trait
Openness to experience	Creative, insightful, intellectually curious
Conscientiousness	Organized, dependable, hardworking
Extraversion	Outgoing, sociable, energetic
Agreeableness	Considerate, cooperative, friendly
Neuroticism	Anxious, irritable, moody

Trait theory suggests that personality is the result of a combination of traits that are relatively stable over time. However, trait theory does not account for external influences on personality or for the reasons underlying personality traits. Furthermore, trait theories are not very successful at predicting specific behaviors because they do not account for the impact of situational factors on behavior. For example, an introvert who generally avoids parties would be more likely to attend his wife's retirement party (ie, specific behavior).

Chapter 26: Personality Theories

Lesson 27.1
The Principles and Components of Stress

27.1.01 Appraisal Theory

Appraisal theory states that one's evaluation (ie, appraisal) of a stimulus determines one's emotional response (see Figure 27.1). In a primary appraisal, an individual classifies a stimulus as threatening, positive, or irrelevant. For a stimulus deemed threatening (ie, a stressor), a secondary appraisal occurs in which the individual evaluates whether their resources/abilities are sufficient to cope with the stressor.

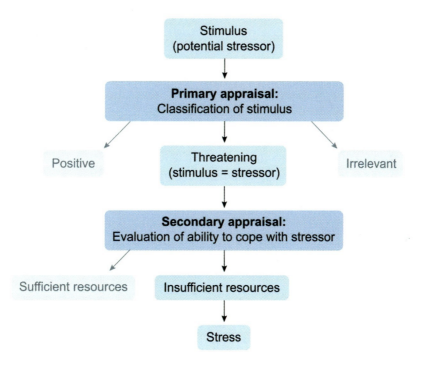

Figure 27.1 Appraisal theory.

27.1.02 Types of Stressors

Stress refers to a threatening or demanding stimulus that disturbs equilibrium in some way (eg, mentally, emotionally, physically). Stress can lead to high blood pressure, anxiety, headaches, and many other health problems. There are four major types of stressors (Figure 27.2):

- Daily hassles are common, everyday occurrences that affect few people and are irritating but are not major stressors (eg, driving in traffic).
- A personal life event is a major life transition that affects few people but is very stressful. Personal life events can be positive (eg, getting married) or negative (eg, a death in the family).
- Environmental (or ambient) stressors are large-scale (ie, affecting many people), minor, but persistent irritations (eg, pollution).
- Catastrophes are large-scale major events that affect many people (eg, natural disasters).

Chapter 27: Stress

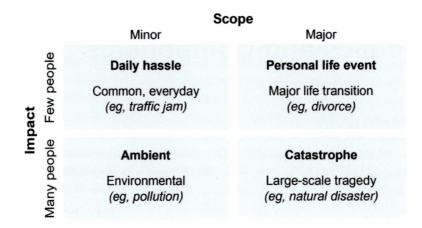

Figure 27.2 Types of stressors.

Stress can result from the complex choices a person may face. Motivational conflict theory describes three inner conflicts regarding choices (Figure 27.3):

- **Approach-approach conflict** occurs when one decides between pursuing two incompatible goals that both have desirable outcomes (eg, going to a party or a movie).
- **Approach-avoidance conflict** occurs when one decides whether to pursue a goal that has both wanted and unwanted outcomes (eg, going on a trip is fun but also expensive).
- **Avoidance-avoidance conflict** occurs when one decides between two alternatives that both have unwanted outcomes (eg, doing chores or getting in trouble for not doing them).

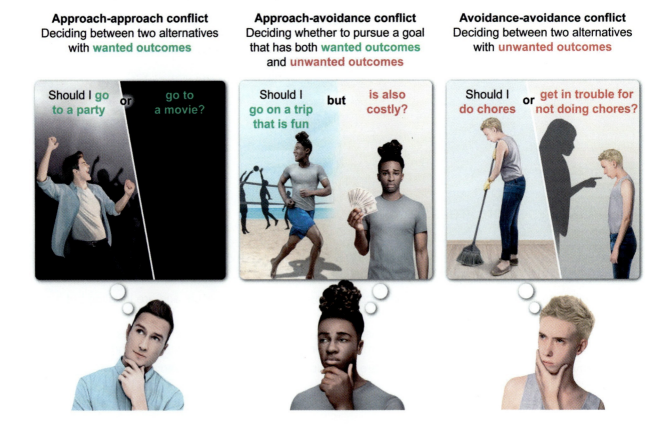

Figure 27.3 Motivational conflict theory.

Lesson 27.2

The Effects of Stress and Stress Management

27.2.01 Effects of Stress

Physiological Response

Immediately after a stressor is detected, the sympathetic ("fight or flight") division of the autonomic nervous system (Concept 4.1.02) mobilizes the body for action. Sympathetic neurons stimulate the adrenal glands (endocrine organs on top of the kidneys) to secrete **norepinephrine** and **epinephrine** (Concept 4.2.03), which helps activate the body to deal with the stressor by increasing cardiac and respiratory activity, increasing blood flow to the skeletal muscles and decreasing digestive functions.

Further, during stress, the hypothalamus causes the pituitary gland to release adrenocorticotropic hormone (ACTH), a stress hormone. ACTH travels in the bloodstream to the adrenal glands, where it causes the release of the stress hormone cortisol that helps prepare the body for action by mobilizing energy stores and modifying immune and metabolic responses.

The general adaptation syndrome (GAS), proposed by Hans Selye, is a model describing how the body reacts to stress. According to this model, there are three stages of the stress response (Figure 27.4):

- The **alarm** stage occurs during the first few minutes of the stress response when a stressor triggers the sympathetic nervous system's fight-or-flight response (eg, increasing the heart rate and perspiration).
- The **resistance** stage can last for hours (eg, exercise), days (eg, final exams), or months (eg, preparing for a graduate school entrance exam). During this stage, the body attempts to establish a new equilibrium in response to an ongoing stressor.
- The **exhaustion** stage occurs if the stressor continues. Prolonged stress depletes energy and results in the body being more vulnerable to negative health effects (eg, depression, viral illness).

	Characteristics	Example
Stage 1: Alarm	A stressor triggers the sympathetic nervous system's fight-or-flight response.	 A student realizes how difficult it will be to keep up with their classes.
Stage 2: Resistance	The body attempts to establish a new equilibrium.	 The student follows a rigid study plan for weeks.
Stage 3: Exhaustion	Energy is depleted, and the body is more vulnerable to negative health effects.	 The student burns out and is unable to keep up with their classes.

Figure 27.4 Hans Selye's general adaptation syndrome.

Emotional and Behavioral Responses

Chronic stress can result in alterations to brain areas that are involved in behavior and emotion. The synaptic changes that occur in the hippocampus as a result of stress impair learning. In addition, chronic stress causes changes in the morphology and functioning of the amygdala and prefrontal cortex. For example, repeatedly stressed rats display behaviors consistent with increased anxiety.

27.2.02 Managing Stress

Research indicates that relaxation, a physiological state of reduced arousal and sympathetic activity, is effective for reducing stress. Relaxation can be achieved through many different techniques (eg, massage, meditation).

Meditation is a practice whereby individuals regulate their awareness and attention with the goal of achieving mental clarity and emotional calmness. Although meditation techniques vary, they can include focusing on one's own breathing or other present-moment stimuli (eg, the sound of rain), or mentally repeating a mantra (a short saying or thought).

Neuroimaging studies have shown that meditation not only results in relaxation but also produces an altered state of consciousness, demonstrated by an increase in alpha brain waves. Regular meditation provides benefits outside of the meditative state, such as improved attention and emotional self-regulation.

Lesson 28.1

Theories of Attitude and Behavior Change

28.1.01 Elaboration Likelihood Model

As Concept 25.1.01 introduces, persuasion involves convincing people to change their attitudes or behavior. The **elaboration likelihood model of persuasion** (Figure 28.1) describes two routes by which a message can cause attitude change in the receiver (the person for whom the message was intended):

- The **peripheral route** to persuasion uses superficial tactics (eg, attractive spokesmodel) to influence attitudes or behaviors.
- The **central route** seeks attitude change after careful consideration of the content of a message. Using the central route to persuade someone would involve making a well-reasoned argument based on facts and evidence.

The central route to persuasion
uses well-reasoned arguments
based on facts and evidence

The peripheral route to persuasion
uses superficial tactics

For example, an individual running for student body president...

...carefully describes the details of her
platform during the school debates
(ie, well-reasoned argument).

...gets the cheerleading squad to perform
a catchy routine in support of her
(ie, superficial tactics).

Figure 28.1 Elaboration likelihood model example.

Either route can result in persuasion, but the central route is more effective when people are willing and able to pay attention to the facts, whereas the peripheral route is more effective when people are not paying close attention to the content of the message.

28.1.02 Social Cognitive Theory

According to **social cognitive theory**, people's behavior and attitudes are learned through **vicarious learning**: observing a model (someone else, such as a friend) engage in a behavior and receive consequences for that behavior. Depending on the outcome, the observer may replicate or avoid the behavior of the model. For example, a soccer player is less likely to commit a foul after watching a teammate get penalized by the referee for committing the same foul.

END-OF-UNIT MCAT PRACTICE

Congratulations on completing **Unit 5: Motivation, Emotion, Attitudes, Personality, and Stress**.

Now you are ready to dive into MCAT-level practice tests. At UWorld, we believe students will be fully prepared to ace the MCAT when they practice with high-quality questions in a realistic testing environment.

The UWorld Qbank will test you on questions that are fully representative of the AAMC MCAT syllabus. In addition, our MCAT-like questions are accompanied by in-depth explanations with exceptional visual aids that will help you better retain difficult MCAT concepts.

TO START YOUR MCAT PRACTICE, PROCEED AS FOLLOWS:

1) Sign up to purchase the UWorld MCAT Qbank
   IMPORTANT: You already have access if you purchased a bundled subscription.
2) Log in to your UWorld MCAT account
3) Access the MCAT Qbank section
4) Select this unit in the Qbank
5) Create a custom practice test

Unit 6 Psychological Disorders and Treatment

Chapter 29 Psychological Disorders

29.1 Types of Psychological Disorders

29.1.01	Anxiety Disorders	
29.1.02	Obsessive-Compulsive Disorder	
29.1.03	Posttraumatic Stress Disorder	
29.1.04	Bipolar Disorder	
29.1.05	Depressive Disorders	
29.1.06	Schizophrenia	
29.1.07	Dissociative Disorders	

29.2 Neurological Disorders

    29.2.01   Parkinson's Disease

Chapter 30 The Biological Underpinnings of Psychological and Neurological Disorders

30.1 The Biological Underpinnings of Psychological Disorders

    30.1.01   The Biological Underpinnings of Schizophrenia
    30.1.02   The Biological Underpinnings of Depression

30.2 The Biological Underpinnings of Neurological Disorders

    30.2.01   The Biological Underpinnings of Parkinson's Disease

Chapter 31 Treatment of Psychological and Neurological Disorders

31.1 Treatment Approaches and Techniques

    31.1.01   Treatment Approaches and Techniques

Lesson 29.1

Types of Psychological Disorders

29.1.01 Anxiety Disorders

Anxiety disorders are characterized by excessive, uncontrollable fear (about a perceived imminent threat) or worry (about a perceived future threat) that impairs functioning.

Generalized Anxiety Disorder

Excessive and uncontrollable worry about a range of events (eg, finances, relationships) for more than six months characterizes **generalized anxiety disorder** (GAD). Individuals with GAD anticipate disastrous outcomes for daily events and activities and find it difficult to control their worry. Muscle tension, difficulty concentrating or sleeping, and feeling restless, fatigued, or irritable are all common symptoms of GAD.

Panic Disorder

Panic attacks, which can be associated with several different disorders (eg, panic disorder, posttraumatic stress disorder) are overwhelming surges of anxiety that peak within minutes. A person who is having a panic attack may experience a racing heart, sweating, chills, trembling, breathing difficulties, dizziness, and/or nausea. **Panic disorder** is characterized by repeated, uncontrollable, and unpredictable panic attacks.

Agoraphobia

Individuals with **agoraphobia** have an intense fear of being unable to escape settings (eg, crowds, public transportation) that cause feelings of panic or being trapped. As a result, individuals with agoraphobia avoid such settings and are often afraid to leave their home. Many individuals with agoraphobia experience panic attacks and/or have panic disorder.

Social Anxiety Disorder

An intense fear of interpersonal rejection or humiliation characterizes **social anxiety disorder**. Individuals with social anxiety disorder become extremely anxious in social settings (eg, at school) and may avoid these situations (eg, public speaking, asking someone on a date).

Specific Phobia

Specific phobia is characterized by excessive, irrational fear of a specific situation (eg, hiking) or animal/object (eg, bears). Some specific phobias are hypothesized to result from the classical conditioning of fear (Concept 17.1.03) through pairing a negative experience (eg, being charged by a bear while out hiking) with a specific animal/object (eg, a bear) or situation (eg, hiking).

29.1.02 Obsessive-Compulsive Disorder

Obsessive-compulsive disorder (OCD) is characterized by obsessions and/or compulsions (see Figure 29.1):

- Obsessions are recurrent, intrusive, distressing thoughts (eg, "someone is going to harm my child").
- Compulsions are repetitive behaviors or rituals (eg, extensive cleaning, counting, organizing behaviors) that are often intended to neutralize obsessions.

The obsessions and/or compulsions must occupy significant time or interfere with functioning for a diagnosis of OCD.

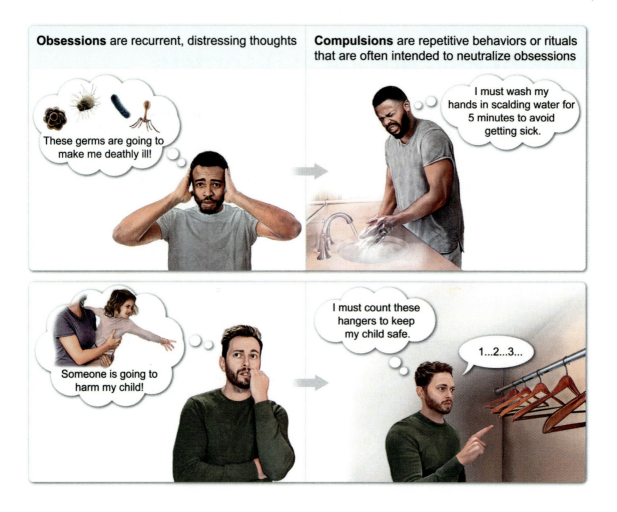

Figure 29.1 Obsessive-compulsive disorder.

29.1.03 Posttraumatic Stress Disorder

Posttraumatic stress disorder (PTSD) arises from exposure to trauma, an event that resulted or almost resulted in death or serious injury (eg, assault, serious accident, combat). PTSD is characterized by reexperiencing the traumatic event (eg, nightmares, flashbacks), negative thoughts and mood (eg, irritability), avoidance of trauma reminders (eg, avoiding certain events, objects, or people), and hyperarousal (eg, hypervigilance, exaggerated startle response, insomnia, difficulty concentrating). For a diagnosis of PTSD, symptoms must be present for at least one month.

29.1.04 Bipolar Disorder

Bipolar I disorder is characterized by mania, defined as abnormally elevated or irritable mood and increased energy. During a manic episode, individuals may be talkative, distractible, overconfident, or act impulsively (eg, drive recklessly). They may feel they need less sleep (eg, staying up for several days). To qualify as a manic episode, symptoms must either require hospitalization or persist for at least one week and be severe enough to impair functioning.

Many people who have bipolar I disorder also experience major depressive episodes. Major depressive episodes share the same symptoms as major depressive disorder (see Concept 29.1.05), such as feelings of sadness or hopelessness, lack of interest in activities, sleep and appetite disturbances, and frequent thoughts of death or suicide.

29.1.05 Depressive Disorders

Depressive disorders are characterized by enduring periods of sadness that interfere with a person's functioning.

One such disorder, **major depressive disorder** (MDD), is characterized by an ongoing period (ie, at least two weeks, but often longer) of depressed mood (severe sadness, hopelessness, or emptiness) and/or a lack of pleasure (anhedonia) or loss of interest in activities. Other symptoms of MDD include changes in appetite (eating more or less); changes in sleep (sleeping more or less); fatigue; low self-worth or guilt; cognitive difficulties (problems with concentration, memory, decision-making); and/or thoughts about death or suicidal ideation. Most people with MDD experience recurrent depressive episodes.

29.1.06 Schizophrenia

Schizophrenia spectrum and other psychotic disorders are severe psychological disorders characterized by a loss of contact with reality.

The most common psychotic disorder is schizophrenia. A diagnosis of **schizophrenia** requires the presence of several characteristic symptoms for one month or longer, at least one of which must be hallucinations, delusions, or disorganized speech. Schizophrenia typically involves both positive symptoms, which are "pathological excesses" (eg, hallucinations, delusions, disorganized speech), and negative symptoms, which are "pathological deficits" (eg, apathy, inability to experience pleasure).

Delusions are fixed, false beliefs that are maintained despite evidence to the contrary (eg, believing that one's private thoughts are being broadcast to others), whereas **hallucinations** are false perceptual experiences in the absence of sensory stimulation (eg, seeing things no one else can).

Many people with schizophrenia also experience psychomotor symptoms (ie, changes in muscle tone or activity), which can occur either as a symptom of schizophrenia or as a side effect of medication.

29.1.07 Dissociative Disorders

Dissociative disorders involve disruptions to memory, consciousness, and/or identity that stem from psychological origins (eg, psychological trauma) as opposed to other medical causes (eg, head injury).

For example, after experiencing a trauma, someone with **dissociative amnesia** might forget important autobiographical details, such as their name or marital status. In rare cases, individuals with dissociative amnesia travel from their home and assume a new identity, which is known as a *dissociative fugue*.

Another dissociative disorder, **dissociative identity disorder**, is characterized by a person having two or more distinct personalities and the inability to recall important autobiographical information (ie, dissociative amnesia).

Lesson 29.2
Neurological Disorders

29.2.01 Parkinson's Disease

Parkinson's disease (PD) is a progressive neurodegenerative disease that is marked by the loss of dopaminergic neurons in the brain's substantia nigra (Figure 29.2). Because these neurons are involved in motor movements, PD results in motor abnormalities: resting tremor (often in the upper extremities), muscle rigidity, slowed movement (ie, bradykinesia), postural instability, and shuffling gait.

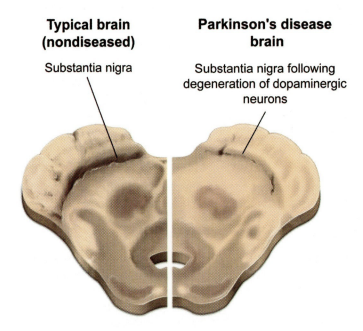

Figure 29.2 Substantia nigra degeneration in Parkinson's disease.

PD can be accompanied by non-motor symptoms including cognitive impairment (eg, problems with memory, executive functioning, visuospatial abilities), sleep disturbances, and depressed and/or anxious mood.

Lesson 30.1

The Biological Underpinnings of Psychological Disorders

30.1.01 The Biological Underpinnings of Schizophrenia

As Concept 29.1.06 discusses, schizophrenia spectrum and other psychotic disorders (eg, schizophrenia) are severe psychological disorders characterized by a loss of contact with reality.

While schizophrenia has a genetic component, there is also evidence that certain environmental factors (eg, prenatal viruses) increase one's risk for developing the disorder. Additionally, there are neurological abnormalities associated with schizophrenia, including enlarged ventricles (spaces in the brain containing cerebrospinal fluid) and decreases in the surrounding brain tissue.

Dopamine has been suggested to play a role in schizophrenia because elevations in the neurotransmitter in certain brain areas have been linked with psychotic symptoms (eg, hallucinations), such as those seen in schizophrenia. Further, many antipsychotic drugs are dopamine antagonists that work in part by blocking the action of dopamine. These observations support the dopamine hypothesis of schizophrenia, which directly attributes the disorder's psychotic symptoms to the activity of dopaminergic neurons.

30.1.02 The Biological Underpinnings of Depression

As Concept 29.1.05 states, depressive disorders (eg, major depressive disorder) are characterized by enduring periods of sadness that interfere with a person's functioning. Major depressive disorder has environmental (eg, abuse during childhood) as well as genetic contributing factors.

Two monoaminergic neurotransmitters (Concept 4.2.03), **serotonin** and **norepinephrine**, have been implicated in depressive disorders because decreases in these neurotransmitters cause symptoms of depression. Further, medications that elevate levels of serotonin and/or norepinephrine alleviate depressive symptoms for many individuals. These observations support the monoamine hypothesis, which directly attributes depression to a deficit in central serotonergic and/or noradrenergic activity.

Some antidepressant medications are agonists of one or more of the monoamines. For instance, selective serotonin reuptake inhibitors (SSRIs) block the reabsorption of serotonin into the presynaptic neuron (ie, reuptake), prolonging the presence of serotonin in the synapse and thereby increasing its action. Another class of medication, monoamine oxidase inhibitors (MAOIs), inhibit monoamine oxidase, an enzyme that contributes to the breakdown of monoamines. The inhibition of monoamine oxidase increases monoamine concentrations, which increases their action.

Lesson 30.2

The Biological Underpinnings of Neurological Disorders

30.2.01 The Biological Underpinnings of Parkinson's Disease

As Concept 29.2.01 introduces, Parkinson's disease (PD) involves the loss of dopaminergic neurons in the **substantia nigra** (SN), which is a brain structure located in the midbrain. The SN projects axons to the **basal ganglia**, a cluster of nuclei that plays an important role in voluntary movements (Figure 30.1).

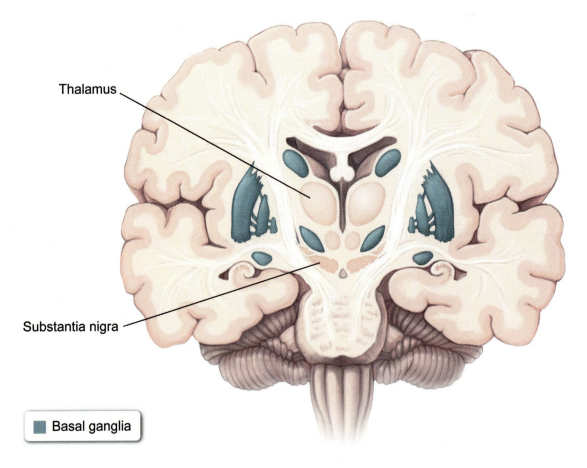

Figure 30.1 Basal ganglia and related structures.

The basal ganglia contributes to the initiation of desired motor programs and also appears to play a role in the inhibition of competing or unwanted movements. The dopaminergic neurons of the SN modulate these actions. The degeneration of the SN in PD impairs this dopaminergic modulation, which causes the motor symptoms of the disease (see Concept 29.2.01).

Some medications used to treat PD (eg, L-dopa) are dopamine agonists, which mimic or enhance the effects of dopamine. Another treatment for PD, deep-brain stimulation (DBS), involves implanting a device in the brain that sends electrical impulses to a specific area. DBS in portions of the basal ganglia has been shown to relieve some of PD's motor symptoms.

Lesson 31.1
Treatment Approaches and Techniques

31.1.01 Treatment Approaches and Techniques

Biomedical versus Biopsychosocial Models

The biomedical model and the biopsychosocial model represent two approaches to conceptualizing how psychological disorders develop and should be treated.

The **biomedical approach** (also known as the biomedical model) to psychological disorders suggests that physiological causes (eg, a deficit in a neurotransmitter system) result in psychological symptoms, and therefore, medical treatment (eg, medication) is advised to fix the underlying problem.

In contrast, the **biopsychosocial model** of psychological disorders suggests that mental disorders are the result of the combination or interaction between biological (eg, genetics, hormones), psychological (eg, thoughts, behaviors), and social (eg, family dynamics, peer groups) factors. Therefore, a biopsychosocial approach to treating a psychological disorder could involve addressing some or all of these factors.

Types of Treatment

Some psychological disorders are most effectively treated with specific medications or therapeutic techniques. For example, research has shown that because schizophrenia is associated with higher levels of dopamine, it is most effectively treated using antipsychotic medication that blocks the action of dopamine (see Concept 30.1.01).

Other psychological disorders are best treated with specific psychotherapeutic techniques. For example, **systematic desensitization** is an effective behavioral (see Lesson 1.2) treatment for specific phobia (see Concept 29.1.01) that pairs relaxation techniques with increasingly distressing stimuli until the client can face the feared animal/object or situation with a diminished fear response.

As depicted in Figure 31.1, a therapist using systematic desensitization could help a client with a specific phobia of rats by initially teaching the client relaxation techniques. The client could then use these skills while facing progressive steps over time, such as by first viewing photos of rats; after some time, holding a life-like toy rat; and eventually (and gradually) approaching a live caged rat.

For example, a therapist using systematic desensitization to help a client with a specific phobia of rats first teaches the client relaxation techniques (eg, breathing exercises)...

...then asks the client to use those techniques with increasingly distressing stimuli until the client is able to face a rat.

Figure 31.1 Example of systematic desensitization.

Major Psychotherapeutic Approaches

More broadly, several major psychotherapeutic approaches have been developed based on theoretical perspectives in psychology.

One major psychotherapeutic approach is **humanistic psychotherapy** (see humanistic theory in Lesson 1.3), which is a person- or client-centered approach that aims to provide a supportive environment in which clients can grow and change. Humanistic therapists demonstrate unconditional positive regard (ie, acceptance and support, regardless of behavior), empathy (ie, seeing the world from the client's perspective), and active listening (ie, paraphrasing a client's own words and asking clarifying questions).

Another major psychotherapeutic approach, **cognitive-behavioral therapy** (CBT), is designed to help individuals replace negative thoughts and behaviors with healthier thoughts and behaviors. For example, a cognitive-behavioral therapist treating an individual with social anxiety (see Concept 29.1.01) could focus on replacing the patient's negative thoughts (eg, "I'm awkward in social situations") and avoidant behaviors (eg, eating alone) with more positive thoughts (eg, "I'm interesting and a great listener") and sociable behaviors (eg, eating with others).

END-OF-UNIT MCAT PRACTICE

Congratulations on completing **Unit 6: Psychological Disorders and Treatment**.

Now you are ready to dive into MCAT-level practice tests. At UWorld, we believe students will be fully prepared to ace the MCAT when they practice with high-quality questions in a realistic testing environment.

The UWorld Qbank will test you on questions that are fully representative of the AAMC MCAT syllabus. In addition, our MCAT-like questions are accompanied by in-depth explanations with exceptional visual aids that will help you better retain difficult MCAT concepts.

TO START YOUR MCAT PRACTICE, PROCEED AS FOLLOWS:

1) Sign up to purchase the UWorld MCAT Qbank
   IMPORTANT: You already have access if you purchased a bundled subscription.
2) Log in to your UWorld MCAT account
3) Access the MCAT Qbank section
4) Select this unit in the Qbank
5) Create a custom practice test

Unit 7 Theories and Research in Sociology

Chapter 32 Theoretical Approaches in Sociology

32.1 Major Approaches in Sociology

    32.1.01   Sociology and Society
    32.1.02   Microsociology and Macrosociology

32.2 Sociological Theories

    32.2.01   Functionalism
    32.2.02   Conflict Theory
    32.2.03   Symbolic Interactionism
    32.2.04   Social Constructionism
    32.2.05   Exchange-Rational Choice
    32.2.06   Feminist Theory

Chapter 33 Research in Sociology

33.1 Empiricism in Sociology

    33.1.01   Empiricism in Sociology

33.2 Types of Studies in Sociology

    33.2.01   Types of Studies in Sociology

Lesson 32.1

Major Approaches in Sociology

32.1.01 Sociology and Society

Sociology is the scientific study of **society** (also known as social life, the social world, or the social environment); society refers to the patterns of relationships and activities developed by a group of people who share a common way of life.

Sociologists examine how humans organize social systems (eg, politics, criminal justice) and create culture (ie, shared languages and customs). When analyzing the relationship between individual actions and the operations of society, sociologists rely on empirical evidence (ie, observations and data) to develop and support their explanations. A few common sociological topics include patterns of group interaction, systems of inequality, and strategies for social change.

32.1.02 Microsociology and Macrosociology

Society is a broad and complicated topic to study. To address this complexity, sociology can be divided into two main approaches, microsociology and macrosociology, which each focus on different aspects of social life.

The **microsociology** (also known as micro-level) approach examines small-scale social phenomena (eg, community events or workplace dynamics) and interpersonal interactions (eg, communication between teacher and student). Microsociology takes a "microscopic view" of society, with a focus on the importance of individual actions that build and shape society. For example, a microsociology approach might focus on communication barriers between physicians and patients to understand the impact of interaction on health outcomes.

The other approach to studying society is the **macrosociology** (also known as the macro-level) approach, which focuses on society-wide institutions (eg, healthcare system) and large-scale events (eg, economic recessions, wars) that impact the everyday experiences of individuals. Macrosociology takes a "bird's-eye view" of society to examine broad patterns, trends, and demographics within the organization of society. For example, a macrosociology approach might focus on how patterns of unemployment and poverty impact the rise of obesity rates.

Both micro- and macro-level approaches are seen as distinct perspectives needed to understand the complexities of social life, and within sociology, neither approach is preferred or considered more accurate. Microsociology examines how individual behaviors impact the larger society, and macrosociology examines how larger social institutions shape the lives of individuals (see Figure 32.1 for a comparison between the approaches).

Focuses on how individual behaviors and interactions build and shape society.

Focuses on how society-wide institutions and large-scale events impact the lives of individuals.

Figure 32.1 Microsociology versus macrosociology.

Lesson 32.2

Sociological Theories

32.2.01 Functionalism

In sociology, a theory is a set of abstract ideas used to explain a wide range of social phenomena such as the development of identity, the emergence of social movements, or the rise of a global economy. Each theory views social life from a particular perspective and is especially relevant for studying certain aspects of society.

One macrosociological theory (Lesson 32.1) is **functionalism** (also known as structural functionalism), which compares society to a biological organism. Émile Durkheim proposed that structures of society (Chapter 43 discusses institutions and systems) work together to maintain stability and order (ie, societal balance), similar to the way various organ systems function to maintain homeostasis in an organism. For example, the healthcare system functions to keep people healthy so that they can work and contribute to society, and the education system functions to teach and train people who will, in turn, contribute to society. Each system serves different purposes and is necessary to support society's overall order and balance.

One application of functionalist theory is to examine different types of social functions (ie, a social institution's purpose or contribution to society). Sociologist Robert Merton argued that social institutions have **manifest functions**, which are expected or planned, and **latent functions**, which are unintended. For example, higher education (ie, social institution) is meant to teach students the skills necessary to become functioning citizens in society (ie, manifest function); however, many students also end up meeting potential romantic partners in college (ie, latent function).

32.2.02 Conflict Theory

Unlike functionalism's focus on balance and order, **conflict theory** views society as a hierarchy of competing groups. According to conflict theory (a macrosociological theory, see Lesson 32.1), tension arises when resources, such as wealth and power (Concept 45.1.05), are unequally distributed throughout society, resulting in the most powerful groups possessing more advantages than others.

The development of conflict theory was influenced by Karl Marx, who wrote about societal changes resulting from the capitalist revolution in the nineteenth century. Marx argued that private ownership (ie, individuals and corporations controlling the means of production instead of the government controlling them) in capitalism would increase inequality between the economic classes. The owning class (ie, **bourgeoise**) gains wealth through the exploitation of the working class (ie, **proletariat**). Marx theorized conflict was inevitable when the working class recognized the inequality within capitalism.

Conflict theory also asserts that power and privilege (Concept 45.1.05) are incorporated within social institutions (eg, economy, criminal justice) through policies and laws. For example, a conflict theory perspective could be applied to understand the inequality in resources for public schools: property taxes fund public education, and this policy results in higher budgets (ie, more resources) for schools in wealthier neighborhoods and lower budgets (ie, fewer resources) for schools in low-income neighborhoods.

32.2.03 Symbolic Interactionism

Symbolic interactionism is a microsociological theory (Lesson 32.1) that views society as the product of social interaction. According to symbolic interactionism, meaning can be communicated through the use of **symbols** (ie, any image, object, gesture, or sound that conveys meaning) within social interaction.

Symbolic meanings differ by context and culture (see Lesson 34.1) and are not permanent. For example, wearing pants was a symbol of masculinity in the United States through the 1950s, but "pants" no longer carry symbolic meaning associated with gender.

George Herbert Mead influenced the development of symbolic interactionism and proposed three key principles:

- Symbolic meaning is created in interaction.
- Individuals act based on the interpreted meaning of symbols.
- Differences in symbolic interpretation result in different actions.

For example, symbols are used in a chemistry lab to communicate different risks and hazards. A chemistry student learns the meaning of these symbols (ie, visual representation of different properties of the chemical) through *interaction* with the instructor and lab assistants. Individuals take precautions when handling the chemicals based on their *interpretation* of the symbols' meaning. An individual who is not a chemistry student may interpret the symbols to have *different meanings* (eg, skull and crossbones could symbolize a pirate ship rather than a toxic chemical).

32.2.04 Social Constructionism

Similar to symbolic interactionism's emphasis on interaction, **social constructionism** is a microsociological theory (Lesson 32.1) focused on the ways societies create ideas and interpret the meaning of reality. A **social construct** is an agreed-upon idea which has been created and supported by a specific social group during a particular time period.

Social constructionism proposes that elements of social "reality" such as common practices, social systems, and identities are social constructs. Objects (eg, wedding ring), behaviors (eg, wearing clothes), and identity categories (eg, nationality) have meaning only because individuals in society have agreed on that meaning. For example, the monetary value of a dollar bill is not based on the physical (ie, inherent) properties of the paper. However, the dollar has value in society because there is agreement on its worth (ie, social construct), and dollar bills are exchanged for goods based on this shared meaning.

32.2.05 Exchange-Rational Choice

Exchange-rational choice refers to two micro-level sociological theories (Lesson 32.1), each offering explanations about human decision-making processes.

The **rational choice theory** proposes that humans are self-interested (ie, take actions that benefit themselves) and make rational (ie, logical) decisions through an analysis of all possible options to maximize gain and minimize loss. For example, a smoker might weigh the costs and benefits of quitting before deciding to act. Rational choice theory can be applied to this example because the smoker's choice about a behavior change is based on a logical analysis of the gains (eg, reduced health risks) and losses (eg, withdrawal symptoms) of smoking cessation, and the decision supports the interests of the individual.

Social exchange theory (also known as exchange theory) applies rational choice theory to interactions and relationships with others. Exchange theory proposes that interactions between people are based on each person's calculation of the benefits (eg, intimacy, support) and costs (eg, time, stress) of the relationship. For example, exchange theory could be useful when studying infidelity in marriages to understand the decision-making process of a spouse leaving or remaining in the relationship.

32.2.06 Feminist Theory

In sociology, **feminist theory** describes a variety of theories aimed to explain the differences in power based on gender (ie, gender inequality). Feminist theories argue that the basis for gender inequality, both historically and in contemporary society, is the organization of society into a patriarchal system (ie, men possess most of the power and prestige as defined in Concept 45.1.05).

Current feminist theory attempts to address how the patriarchal structure of society negatively impacts all individuals, regardless of gender identity. For example, traditional gender roles within the family (ie, father as provider and mother as nurturer) can lead to the development of distant relationships between fathers and children.

Feminist theories include both macro- and micro-level approaches (Lesson 32.1). At the macro-level, feminist theory considers how large-scale social processes maintain gender inequality. For instance, the term *glass ceiling* refers to workplace practices that discriminate against women (eg, less pay, fewer promotions), resulting in the underrepresentation of women in certain fields (eg, surgery) and in leadership positions (eg, CEO).

At the micro-level, feminist theory considers how one-on-one interactions also maintain gender inequality through objectification (ie, positioning women as sexual objects) or devaluation (ie, negative or oppressive language towards women). For example, the derogatory phrase "you throw like a girl" is aimed to devalue femininity at the micro-level of interaction.

Lesson 33.1
Empiricism in Sociology

33.1.01 Empiricism in Sociology

Research in sociology shares many of the same principles as research in psychology, including the scientific method (Lesson 2.1), reliance on credible results (Lesson 2.3), and ethical research practices (Lesson 2.4). Sociologists study society and social interaction through research based on **empiricism** (ie, objective observations) rather than relying on intuition or personal experience.

Lesson 33.2

Types of Studies in Sociology

33.2.01 Types of Studies in Sociology

In sociology, research can be quantitative, qualitative, or a combination of both—referred to as mixed methods. (Note: while psychology also uses quantitative, qualitative, and mixed methods research designs, this lesson focuses on methods in sociology.)

Quantitative research collects large-scale, numeric data that can be analyzed statistically to determine relationships between defined variables (Concept 2.1.02). Types of quantitative research methods include surveys (eg, multiple-choice questionnaires developed for a specific study) and secondary data analysis (eg, research based on existing U.S. Census datasets). For example, statistical analysis of data from a survey administered to pre-med students could determine if there is a relationship between the numeric rank of undergraduate university and students' medical school entrance exam scores.

In contrast, **qualitative** research gathers in-depth, non-numeric data (eg, words, cultural practices as defined in Concept 34.1.01) to analyze the presence of patterns or themes. Types of qualitative research methods include ethnography (ie, intensive observations and interviews of a social group) and focus groups (ie, group interview and discussion on a specific topic or experience). For example, analysis of data from a focus group of pre-med students could identify how social factors such as socioeconomic status (defined in Concept 45.1.01) impact how students prepare for the medical school entrance exam.

Mixed methods research employs both quantitative and qualitative approaches and can be a more comprehensive way to understand a social phenomenon. For example, a researcher could conduct a *quantitative* survey to analyze the relationship between undergraduate university rankings and medical school entrance exam scores as well as a *qualitative* focus group with pre-med students about social factors impacting exam preparation to gain a more complete understanding of the preparation process for the exam.

See Figure 33.1 for a summary of quantitative, qualitative, and mixed methods research designs.

Quantitative
- Data is numeric
- Analysis identifies relationships between variables through statistics

Mixed methods
- Intergrates both quantitative and qualitative approaches
- Can be a more comprehensive way to understand social phenomenon

Qualitative
- Data is non-numeric (eg, words, cultural practices)
- Analysis identifies patterns or themes in the data

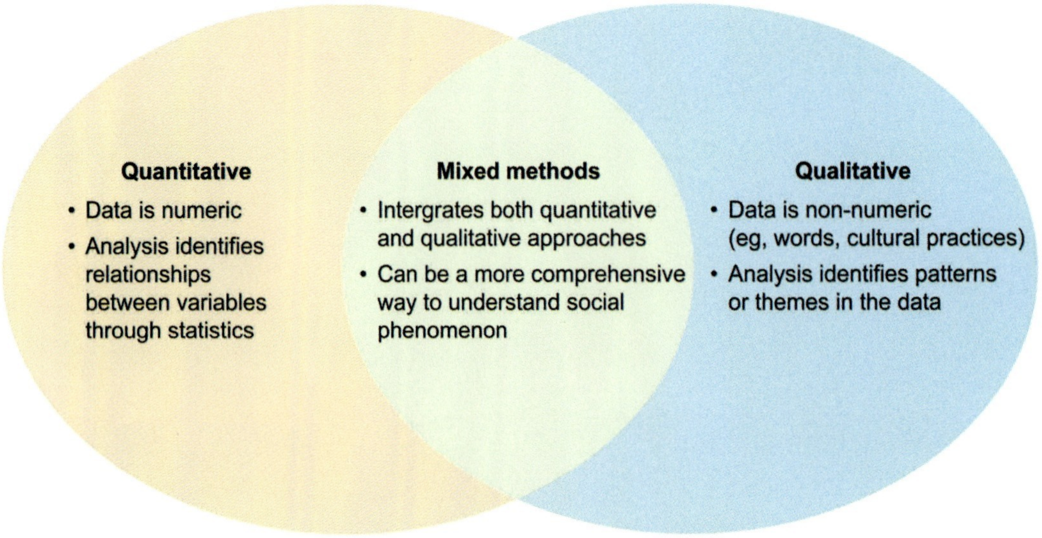

Figure 33.1 Types of research designs.

END-OF-UNIT MCAT PRACTICE

Congratulations on completing **Unit 7: Theories and Research in Sociology**.

Now you are ready to dive into MCAT-level practice tests. At UWorld, we believe students will be fully prepared to ace the MCAT when they practice with high-quality questions in a realistic testing environment.

The UWorld Qbank will test you on questions that are fully representative of the AAMC MCAT syllabus. In addition, our MCAT-like questions are accompanied by in-depth explanations with exceptional visual aids that will help you better retain difficult MCAT concepts.

TO START YOUR MCAT PRACTICE, PROCEED AS FOLLOWS:

1) Sign up to purchase the UWorld MCAT Qbank
   IMPORTANT: You already have access if you purchased a bundled subscription.
2) Log in to your UWorld MCAT account
3) Access the MCAT Qbank section
4) Select this unit in the Qbank
5) Create a custom practice test

Unit 8 Identity and Social Interaction

Chapter 34 Culture

34.1 The Components of Culture
- 34.1.01 The Components of Culture

34.2 Types of Culture
- 34.2.01 Material versus Symbolic Culture
- 34.2.02 Dominant Culture, Subculture, and Counterculture

34.3 Cultural Change
- 34.3.01 Cultural Lag
- 34.3.02 Culture Shock
- 34.3.03 Assimilation and Multiculturalism
- 34.3.04 Mass Media and Popular Culture
- 34.3.05 Transmission and Diffusion

34.4 Socialization
- 34.4.01 Agents of Socialization
- 34.4.02 Primary and Secondary Socialization

Chapter 35 Identities and Identity Formation

35.1 Types of Identities
- 35.1.01 Self-Concept and Social Identity
- 35.1.02 Types of Social Identities

35.2 Identity Formation
- 35.2.01 Theories of Identity Development
- 35.2.02 Influence of Culture and Socialization on Identity Formation
- 35.2.03 Influence of Social Factors on Identity Formation

Chapter 36 Interacting with Others

36.1 The Presentation of Self
- 36.1.01 Dramaturgical Approach
- 36.1.02 Impression Management

36.2 Status
- 36.2.01 Status and Status Set
- 36.2.02 Types of Status

36.3 Roles
- 36.3.01 Roles
- 36.3.02 Role Strain, Role Conflict, and Role Exit

36.4 Groups
- 36.4.01 Primary and Secondary Groups
- 36.4.02 In-Group versus Out-Group
- 36.4.03 Group Size

36.5 Networks

36.5.01	Networks and Social Network Analysis
36.5.02	Social Capital and the Strength of Weak Ties

36.6 Organizations

36.6.01	Types of Organizations
36.6.02	Bureaucracy
36.6.03	McDonaldization

Chapter 37 Attraction, Aggression, and Attachment

37.1 Attraction

37.1.01	Attraction

37.2 Aggression

37.2.01	Aggression

37.3 Attachment

37.3.01	Attachment

Chapter 38 Social Behavior in Animals

38.1 Altruism

38.1.01	Altruism

Chapter 39 Attributing Behavior to Others

39.1 Attributional Processes

39.1.01	Attributional Processes

Chapter 40 Prejudice, Stereotypes, and Discrimination

40.1 Prejudice

40.1.01	The Contributions of Power, Prestige, and Class in Prejudice
40.1.02	The Role of Emotion in Prejudice
40.1.03	The Role of Cognition in Prejudice

40.2 Stereotypes

40.2.01	Stereotype Threat
40.2.02	Self-Fulfilling Prophecy
40.2.03	Stigma
40.2.04	Ethnocentrism and Cultural Relativism

40.3 Discrimination

40.3.01	Prejudice and Discrimination
40.3.02	Individual versus Institutional Discrimination

Chapter 41 Group Processes and Behavior

41.1 Group Processes

41.1.01	Social Facilitation
41.1.02	Social Control

41.2 Social Loafing, the Bystander Effect, and Deindividuation

    41.2.01   Social Loafing
    41.2.02   The Bystander Effect
    41.2.03   Deindividuation

41.3 Conformity and Obedience

    41.3.01   Conformity
    41.3.02   Obedience

41.4 Group Decision-Making

    41.4.01   Group Polarization
    41.4.02   Groupthink

Chapter 42 Normative and Non-normative Behavior

42.1 Social Norms

    42.1.01   Folkways, Mores, and Taboos
    42.1.02   Sanctions
    42.1.03   Anomie

42.2 Deviance

    42.2.01   Deviance
    42.2.02   Perspectives on Deviance

Lesson 34.1
The Components of Culture

34.1.01 The Components of Culture

To live cooperatively, societies develop a **culture**, or way of life, through shared customs and ideas. Culture encompasses many aspects of social life, including physical structures (eg, style of architecture, design of transportation systems) and typical practices (eg, rules of politeness, standards of dress, traditional diet). Individuals learn about their culture through experiences with others (eg, family, community, school), which allows culture to be transferred from one generation to the next. Sociologists have identified common components of culture, including values, beliefs, symbols, language, and rituals.

Values and beliefs are two related concepts used to describe how shared ideas shape social life. **Values** are what a society holds as moral, desirable, or important. **Beliefs** are the ideas about what is true or sacred that help guide human behaviors. For example, in capitalistic societies, the value of wealth is prioritized as a sign of success and a good life. Individuals living in capitalist societies often also believe in meritocracy (ie, individual efforts and skills drive success) and act in ways aimed to achieve wealth as a sign of a successful life (eg, working two jobs to buy a bigger house).

A shared understanding of symbols is needed to communicate meaning within a culture. A **symbol** is anything (eg, object, word, gesture) that represents or stands for something else (eg, a flag represents a country). The meaning of a symbol is specific to a culture and could have a different meaning in another culture. For example, the hand gesture made by forming an "O" with the index finger and thumb represents "okay" in U.S. culture but is an offensive symbol that represents contempt in Brazilian culture.

One complex symbol system central to culture is **language**, which serves as the foundation for interaction in a society. Language includes all spoken, written, and nonverbal communication and relies on symbols to represent ideas. For example, the word "physician" uses letters to symbolize the idea of the medical occupation. A shared language system is also crucial for the transmission of culture (see Concept 34.3.05) through communication about beliefs and customs between members of society.

Cultural practices are the behaviors that members of a society typically engage in, such as maintaining a gendered appearance (eg, certain clothing, hairstyles, grooming) or using customary greetings (eg, handshake, bow). **Rituals** are a type of cultural practice wherein individuals participate in traditional behaviors associated with a specific ceremony (eg, baptisms, bar mitzvahs, funerals) or celebration (eg, birthdays, holidays). For example, a marriage ceremony (depicted in Figure 34.1) is a ritual connected to customs and traditional behaviors that can vary across cultures.

Chapter 34: Culture

A marriage ceremony is a cultural *ritual* connected to the value of family and the *belief* in long-term, committed relationships. Many *symbols* are present including exchanging of rings to represent their union and the *language* of the marriage vows.

Figure 34.1 Diversity in the components of culture in marriage ceremony rituals.

Lesson 34.2
Types of Culture

34.2.01 Material versus Symbolic Culture

Components of culture (see Lesson 34.1) can also be divided into two categories, material and symbolic culture, to differentiate between the objects and the ideas of a society, respectively (outlined in Table 34.1).

Humans depend on tools and technology to sustain a way of life; **material culture** describes the *tangible* artifacts used by society such as toys, dwellings, art, and machines. These physical objects shape the way humans behave. For example, changes in communication technology, such as the shift in popularity of handwritten letters to phones calls, impact the ways humans interact with one another. Examining material culture can also help illustrate what is valued by society. For example, a microwave (ie, material culture) is a human-made tool used to prepare food quickly and reflects the importance of speed (ie, value of efficiency) in contemporary societies.

In contrast, most of the components (ie, values, symbols, beliefs) that Lesson 34.1 describes are a part of **symbolic culture** (also called nonmaterial culture), which refers to the *intangible* elements of a culture. The symbols and beliefs of a culture shape how individuals engage with others. For example, beliefs about aging guide how a society interacts with its elderly members. Symbolic culture communicates shared meaning within society through knowledge systems (eg, symbols/language, religion), traditional stories (eg, historical accounts, folklore), and values (eg, "pursuit of happiness").

Table 34.1 Differences in material and symbolic culture.

	Definition	Medical example
Material culture	Tangible objects valued by society, including clothing, tools, and technology	A stethoscope is an object used to listen to the heart and lungs.
Symbolic culture	Intangible elements of a society used to convey meaning, including symbols, values, and folklore	The caduceus (rod and snake) is a symbol of medical professions.

Although material and symbolic culture are defined as two distinct categories, physical objects can also have symbolic meaning outside of the purpose or function of the object. For example, an automobile is an object of material culture because it is a tangible artifact used for transportation in society. However, the type of car an individual drives also carries symbolic meaning including the status of the brand (eg, economical or luxury), the value of patriotism (eg, international or domestic production), and/or beliefs about the environment (eg, gas-powered or electric).

34.2.02 Dominant Culture, Subculture, and Counterculture

In small communities, the components of culture may be relevant and uniform for all individuals. However, in large societies, multiple cultures often exist together with a diversity of beliefs, symbols, and practices. The concepts of dominant culture, subculture, and counterculture are used to describe types of cultural variation (ie, differences) present within society (see Figure 34.2).

The **dominant culture** includes the traditional set of values, beliefs, and rituals that define a society. Components of the dominant culture are widely accepted as mainstream and reflect the beliefs and practices of the most powerful groups in society. For example, although the United States does not have a national religion, Christianity can be seen as a part of the dominant culture. Symbols and rituals from Christianity are incorporated into other parts of society such as legal oaths taken on the Christian bible and public schools/businesses recognizing Christian celebrations as official holidays (eg, Christmas, Easter).

One type of cultural variation is a **subculture**, which refers to a group of individuals whose values and practices generally align with the dominant culture but who also possess some *distinct* characteristics (eg, ways of speaking, attire, rituals). Members of subcultures tend to identify as a group and engage in activities recognized by others as different from the dominant culture. For example, the military is a subculture in the United States in which the values of the dominant culture (eg, freedom, democracy) are upheld, but the group has a distinct way of life (eg, uniforms, style of communication, power structure).

Another type of cultural variation is **counterculture**, which refers to a subset of society that *opposes* and/or *rejects* the mainstream values and practices of the dominant culture. Sometimes countercultures operate independently and are isolated from mainstream society (eg, the Amish are a religious counterculture living separately from the dominant culture). Other times, countercultures work within a society with the goal of changing the dominant culture (eg, the antiwar movement of the 1960s was a counterculture opposing the Vietnam War, which was supported by the dominant culture).

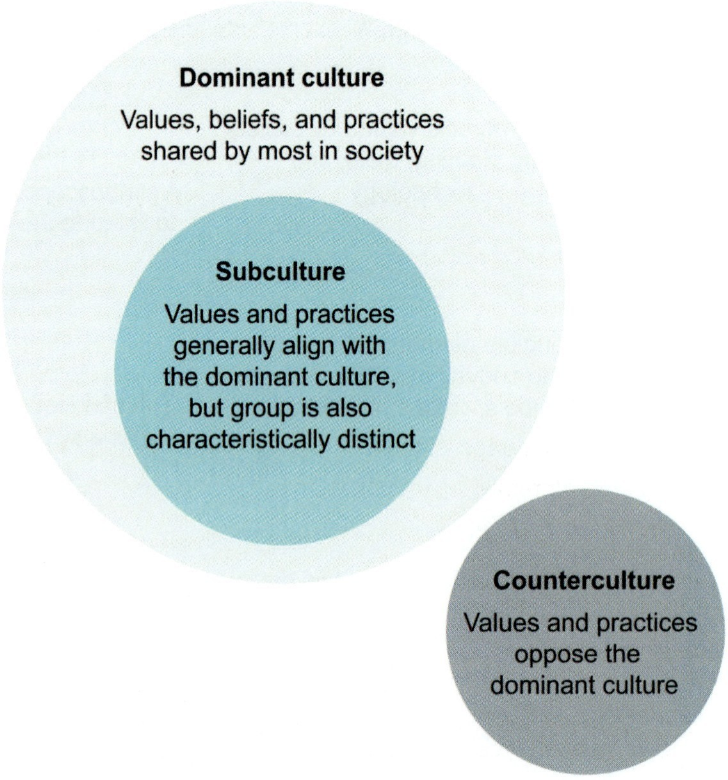

Dominant culture
Values, beliefs, and practices shared by most in society

Subculture
Values and practices generally align with the dominant culture, but group is also characteristically distinct

Counterculture
Values and practices oppose the dominant culture

Figure 34.2 Variations in culture.

Lesson 34.3

Cultural Change

34.3.01 Cultural Lag

Cultural values and behaviors change over time and differ across societies.

Technological innovation can occur quickly without allowing time for societies to make necessary adjustments (eg, the invention of automobiles came before the development of road safety laws). **Cultural lag** describes the time delay between rapid changes in material culture and slower changes in symbolic culture (see Concept 34.2.01 on material and symbolic culture), which can create social problems (eg, unsafe driving conditions before the implementation of traffic laws).

For example, artificial intelligence (AI) is a rapidly developing technology impacting many parts of society, including medicine (eg, symptom checker chatbots). Cultural beliefs about the appropriate use of AI technology are slower to change (ie, cultural lag), and many individuals question the consequences (eg, misdiagnosis, insecure patient data) of replacing human interaction with AI.

34.3.02 Culture Shock

Each society has its own customs and routines of everyday life. When people travel or move to a new country, they may experience challenges while immersed in a different society. **Culture shock** (Table 34.2) involves feelings of disorientation, uneasiness, and even fear associated with the unknown culture. The causes of culture shock largely result from difficulty communicating due to language barriers and unfamiliarity with symbols, values, and/or customs. For example, differences in food customs (such as traditional diet, eating etiquette, timing of meals) may cause culture shock as an individual is exposed to a new culture and adjusts to that way of life.

Table 34.2 Elements of culture shock.

Characteristics	*Unpleasant emotions associated with new culture include:* • Disorientation • Anxiety • Fear
Causes	*Difficulty communicating and understanding due to:* • Language barriers • Unfamiliar symbols, signs • Different norms, values

34.3.03 Assimilation and Multiculturalism

Geographic mobility (ie, moving between cities or regions within a country) and immigration (ie, moving from one country to another) create the presence of multiple cultures within the same society. To address cultural differences, societies engage in informal practices and develop formal policies or laws that either minimize or maintain cultural distinctions.

One approach to cultural differences is **assimilation**, which is the forced or voluntary process of cultural integration in which people adopt the values, symbols, and rituals of the dominant culture. Through assimilation, immigrant and/or subculture groups begin to resemble mainstream society, which reduces cultural distinctions. For example, if Indian immigrants who live in the United States replace their practice of Ayurveda (traditional Indian medicine) with biomedical (Western) treatments to cure an illness, they are engaging in assimilation.

A contrasting approach to addressing cultural differences is **multiculturalism** (see Figure 34.3). Unlike assimilation, which results in adaptation to mainstream society, multiculturalism promotes the recognition and accommodation of cultures that differ from the dominant culture (eg, posting airport signs in multiple languages). The outcome of multiculturalism is a diverse society that advocates for the respect and protection of various cultures to coexist. For example, the variety of cultural clubs and activities on many college campuses reflects multiculturalism by encouraging the appreciation of cultural differences.

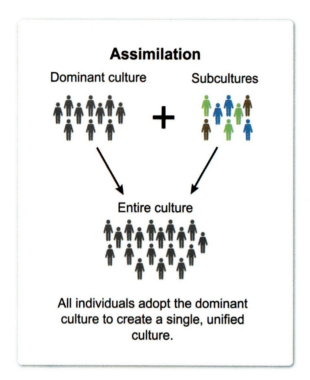
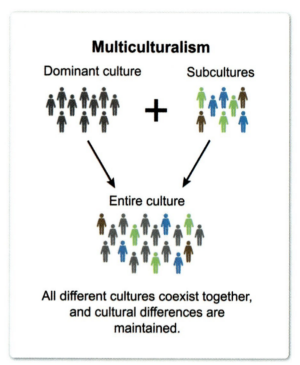

Figure 34.3 Assimilation versus multiculturalism.

34.3.04 Mass Media and Popular Culture

Cultural *differences* are addressed by assimilation or multiculturalism, but sociologists also examine the cultural practices *common* to many people. **Popular culture** refers to the beliefs, trends, and behaviors that are widespread and relevant in a society. For example, in U.S. society, activities such as wearing jeans, watching superhero movies, and reading best-seller books are considered features of popular culture.

Popular cultural trends and behaviors are often spread through **mass media** (eg, television, newspapers, the Internet), the social institution responsible for the communication of information within a society. For example, the promotion of a newly released movie on TV talk shows, commercials, and online platforms illustrates how mass media creates widespread appeal for the movie to establish a trend in popular culture.

34.3.05 Transmission and Diffusion

Culture can be transferred within and between societies (see Table 34.3). **Cultural transmission**, which describes the passing of cultural elements from one generation to the next, serves to stabilize the beliefs and behaviors of a society across time. Components of culture are transmitted within a society through education and socialization (Lesson 34.4 further explains this process), such as learning the national anthem in school.

In addition, **cultural diffusion**, which is the spread of cultural elements from one society to another, expands the scope of culture through interaction between different societies. For example, the introduction of sushi (Japanese cuisine) in the United States illustrates the global spread of culture in which food customs are no longer tied to a specific group.

Table 34.3 Cultural transmission and diffusion.

	Definition	Example
Cultural transmission	Passing on beliefs and behaviors from an older generation to a younger generation	A child learns how to cook traditional cuisine from a parent.
Cultural diffusion	Spreading of culture from one group to another	People in the United States celebrate Cinco de Mayo.

Lesson 34.4

Socialization

34.4.01 Agents of Socialization

Humans are social beings who typically live cooperatively, but to effectively interact with others, the specific components of how a society is organized and operates must be learned. **Socialization** is the mechanism societies use to pass elements of culture from one generation to the next. Through interactions and experiences with others, an individual gains cultural skills and knowledge required to be a member of society.

Individuals experience significant interactions with **agents of socialization** that include the people, groups, and institutions who actively facilitate learning about social life (see Table 34.4). Relationships with family and/or caregivers are formed early in life and provide guidance on fundamental elements of culture such as typical behaviors, language, and core beliefs. Friends and peers often have similar life experiences and serve to either reinforce or contradict the values and practices taught by family.

Schools and the workplace are institutions that also contribute to socialization by teaching standards of behavior (eg, working quietly) and cultural values (eg, respect for authority). The consumption of mass media content serves as another agent of socialization by shaping common knowledge, attitudes, and trends within a society. Recent changes in mass media, such as the proliferation of social media platforms, also provide new means of communication with other agents of socialization, such as family and friends.

Table 34.4 Agents of socialization.

Family	• High degree of contact, especially early in life • Provides earliest socialization, teaching norms, values, behaviors, etc.
Friends/peers	• Usually similar age and life circumstances (eg, same neighborhood) • High degree of contact, especially during adolescence • May socialize in ways that confirm or contradict family values
School/workplace	• Typically brings together many different types of people • Standardized behaviors are taught and usually enforced (eg, arriving on time)
Mass media	• Social institution that spreads information • Helps shape attitudes and change cultural norms and values • Provides new means of communication (eg, social media)

34.4.02 Primary and Secondary Socialization

Socialization is an ongoing process that operates differently at various stages of the life course. **Primary socialization** refers to early childhood experiences in which individuals learn the necessary skills (eg, language) and worldviews (eg, cultural values, religious beliefs) to effectively participate in society. Family, peers, and schools are the common agents of socialization during this intense and extensive stage.

Socialization also impacts how individuals adapt to changes in social positions. **Secondary socialization** occurs later in life when an individual learns to be a part of a new group or subculture (eg, medical school students learn the professional code of conduct for physicians).

One type of secondary socialization is **anticipatory socialization** in which individuals prepare for changes in expected behaviors, such as attending a parenting class before having a child. Another type is **resocialization**, which occurs when an individual enters a new social environment, such as when joining the military, and must learn new cultural skills and behaviors.

Lesson 35.1
Types of Identities

35.1.01 Self-Concept and Social Identity

Within the behavioral sciences, **identity** refers to the way a person perceives themselves with regard to individual characteristics (eg, personal abilities such as singing), life experiences (eg, parenthood), and group membership (eg, belonging to a religious community). Two related yet distinct aspects of identity are self-concept and social identity.

The way an individual thinks and feels about themselves is referred to as **self-concept** (ie, how to answer the question "Who am I?"). Humanistic psychologist Carl Rogers posited that self-concept is a primary part of personality (see Lesson 26.2).

On the other hand, **social identity** (also known as the social self or sense of self) describes how individuals perceive themselves as members of social groups (see Lesson 36.4). Social identities define individuals in relation to others and allow for social groupings around shared qualities and experiences such as family, occupations, and organizations. The identity and characteristics of the group can become incorporated into an individual's sense of self. For example, an individual may begin to see themselves as a "Christian" who is devout and faithful (ie, group characteristics) after experiences within a Christian church community.

Individuals have multiple social identities (eg, parent, woman, surgeon), and the social situation helps determine the most relevant social identities. This allows for social identities to be flexible with the ability to change. The concept of **salience** describes the process of determining which identity is most important in the current social context. For example, the identity of "student" is salient (ie, most important) while attending medical school, but the identity of "physician" would become more important during residency.

35.1.02 Types of Social Identities

Sociologists recognize different types of identities (also known as **social identity categories**) that have greater significance in society, including race/ethnicity, gender, age, sexual orientation, and class (see descriptions in Chapter 44). Like the impact of groups on social identity formation, common types of identities help provide a framework to develop a sense of self. For example, if an individual lives in a society where there is prejudice (Lesson 40.1) toward older adults (eg, older people are "difficult to deal with" and "less physically capable"), he may incorporate these negative beliefs into his sense of self as he ages.

Social identity categories also help sociologists understand the patterns of experiences (see Chapter 44 on demographics) and systems of inequalities (Lesson 45.1) in society. For example, a sociologist could use types of identities to study the patterns of experience for women working in STEM (ie, science, technology, engineering, and mathematics) professions, which, historically, have been male-dominated fields.

Lesson 35.2
Identity Formation

35.2.01 Theories of Identity Development

A number of psychological theories describe how certain aspects of an individual (eg, personality, cognition, morality) develop across the lifespan. Some of these theories are discussed elsewhere in this book; for example, Jean Piaget's theory of cognitive development is discussed in Lesson 20.1 on cognition. Two additional developmental theories, Erik Erikson's psychosocial theory and Lawrence Kohlberg's stages of moral development, are also important to consider in the context of identity formation.

Each of the eight stages in **Erik Erikson's psychosocial theory** involves an age-related crisis or conflict, which is an opportunity for individual growth and social development. Resolution does not necessarily occur in each stage before an individual moves on to the next. However, Erikson believed that unresolved conflict forms the basis for adult psychopathology and other maladaptive behaviors. In order, the stages across the lifespan are:

1. **Trust versus mistrust**: Infants (0–1 year) with sensitive, attentive caregivers will develop a sense of trust; those with inconsistent care will not.
2. **Autonomy versus shame/doubt**: Toddlers (1–3 years) who are encouraged will develop independence; those who are scolded for failure will feel shame and doubt.
3. **Initiative versus guilt**: Children (3–6 years) who successfully interact with others will develop a sense of initiative; those who are criticized will experience guilt.
4. **Industry versus inferiority**: Children (6–12 years) who successfully develop new skills will feel industrious; those who are not encouraged will feel inferior.
5. **Identity versus role confusion**: Adolescents (12–20 years) who successfully interact with peers will develop a sense of self-identity; those who do not will experience role confusion.
6. **Intimacy versus isolation**: Adults (20–40 years) who can commit to and love others will develop a sense of intimacy; those who cannot will feel isolated.
7. **Generativity versus stagnation**: Adults (40–65 years) who successfully contribute to society will feel productive; those who do not will feel stagnant.
8. **Integrity versus despair**: Older adults (>65 years) who feel accomplished will gain a sense of integrity; those who do not will feel depressed and hopeless.

Another theory, **Lawrence Kohlberg's stages of moral development**, focuses on moral and ethical reasoning. Kohlberg's theory was developed from studies in which individuals were asked to respond to moral dilemmas. For example, Kohlberg asked participants to consider a scenario in which a man was unable to afford a medication that would save his wife's life. Participants were asked whether the man should steal the medication and, more importantly, *why* they thought he should (or should not) steal the medication.

Based on the results, Kohlberg developed a model of moral development characterized by six stages (summarized in Table 35.1). At the **pre-conventional level**, morality is controlled by outside forces: individuals attempt to avoid punishment (Stage 1) and try to maximize an exchange of favors with others (Stage 2). At the **conventional level**, morality is defined by existing social norms and values: individuals want to be "good" and liked by others (Stage 3) and obey laws (Stage 4). At the **post-conventional level**, morality is based on universal moral principles: laws are viewed as flexible, considered in the context of helping the greatest number of people (Stage 5), and justice and human dignity apply universally to all people (Stage 6).

Although Kohlberg stated that an individual progresses through the stages in sequence, he did not assign age ranges to the stages. Furthermore, Kohlberg asserted that most adults do not progress past Stages 3 or 4.

Table 35.1 Lawrence Kohlberg's stages of moral development.

Level	Defined by	Stage	Moral reasoning
Pre-conventional	*Direct consequences to the individual*	1. Obedience	Avoiding punishment by authority *(eg, I'm not going to steal because I'll get in trouble)*
		2. Self-interest	Expecting equal exchange to further own self-interest *(eg, I'll help you if you help me)*
Conventional	*Society's norms and values*	3. Conformity	Wanting to be "good" to secure the approval of others *(eg, I'll do my homework so the teacher likes me)*
		4. Law and order	Obeying laws of society *(eg, I'm not going to speed because it's against the law)*
Post-conventional	*Own ethical principles*	5. Social contract	Maximizing benefit for the largest number of people *(eg, it's okay to break a law if it saves a life)*
		6. Universal ethical principles	Following own ethical principles of justice above all else *(eg, I take action against laws violating basic human rights)*

35.2.02 Influence of Culture and Socialization on Identity Formation

In sociology, **identity formation** describes the process of synthesizing and integrating various types of identities (see Concept 35.1.02) into a unified sense of self, which is shaped by socialization and cultural context.

To form a social identity (ie, perception of self in relation to others), an individual must interact within social groups and experience socialization (ie, learning to be a member of the group). An individual's social identity is shaped by interactions with agents of socialization (eg, family, peers, media, which are defined in Lesson 34.4) wherein other people provide information about the self (see Concept 35.2.03 for details on the role of social factors on identity formation).

Every culture (Lesson 34.1) has a specific way of life, including definitions of social roles (eg, gender roles) and cultural practices (eg, rites of passage), which shape how an individual forms their identity. For example, individuals raised in more collectivist (ie, interdependent) cultures such as Japan tend to view their identity in terms of group membership, whereas those raised in more individualistic (ie, independent) cultures such as the United States tend to view their identity in terms of personal characteristics.

35.2.03 Influence of Social Factors on Identity Formation

Sociologist George Herbert Mead suggested that there are two aspects of the social self, which emerge through interaction with significant others (eg, parents, siblings) who play formative roles in an individual's life. The **"I"** refers to the unsocialized, spontaneous self (eg, a child demands a toy instantly even though the parent is on the phone). The **"Me"** is the socialized, reflective self (eg, a child understands they need to wait until the parent is off the phone to ask for a toy).

According to Mead, the "I" and "Me" aspects of the self develop in the following stages (summarized in Table 35.2):

- **Imitation**: Babies/toddlers mimic others (eg, a parent's hand gesture of waving "goodbye") and begin using symbols and language (eg, repeating a phrase used by a parent) without understanding the symbolic meaning of the words or behaviors. At this stage, children have no sense of "self" as separate from the world around them.
- **Play**: Through play (eg, pretending to be a doctor), preschool-age children begin **role-taking** (ie, understanding the perspectives of others). When children understand themselves as individuals separate from others, the "I" component of the self has developed. Children then begin to imagine how others perceive them, which is the beginning of the development of the "Me."
- **Game**: School-age children become aware of their social position in relation to others (eg, understanding rules, positions, and strategies when participating in a game of baseball). They begin to see themselves from the perspective of the more abstract, **generalized other** (ie, broader societal expectations), further developing the "Me" to incorporate the values and rules of the society in which they live.

Table 35.2 George Herbert Mead's theory of the social self.

Stage	Age	Development process	Aspects of self
Imitation	Infancy, toddler	**Imitation**: mimic the behavior of others	No sense of self separate from others
Play	Preschool age	**Role-taking**: taking on roles to understand the perspective of others	"I" is developed and "Me" begins to develop
Game	School age	**Generalized other**: holds an understanding of broader social expectations	"Me" is fully developed

Another theory about the influence of social factors on identity formation is the **looking-glass self**. According to Charles Cooley, interaction serves as a mirror (ie, looking glass) where the way an individual is perceived and treated by others is reflected in how they perceive themself. Identity develops through an individual's interpretation of what others think and feel about them. For example, if a student is viewed and treated as intelligent in most classroom experiences, these interactions can influence the incorporation of "intelligent" into their identity.

Lesson 36.1

The Presentation of Self

36.1.01 Dramaturgical Approach

Sociologist Erving Goffman developed a theory of social interaction that focused on the presentation of the self. Goffman describes social interaction as an exchange of role performances that communicate information about the self to others.

The **dramaturgical approach** is a sociological theory that examines social interaction using a theater metaphor. In this theory, social life is like a play in which individuals are "actors" following a script to perform a role in front of others, the "audience." The social situation and physical location set the "stage" for interaction that guides what social roles are performed and which behaviors are expected. The audience evaluates the success of the social performance much like a theatrical audience can applaud or boo at a play.

The concepts of front-stage and back-stage describe differences in the presentation of the self. The **front-stage** is social interaction involving role performance and audience evaluation. Most social interaction occurs in front-stage settings in which social actors present an idealized self-performance to others. The **back-stage** is social interaction involving informal or relaxed performances without the evaluation from an audience. In the back-stage, interaction with close friends, colleagues, or family does not include the presentation of an idealized self and can also be a space to rehearse and prepare for a front-stage performance.

A typical restaurant exemplifies front-stage and back-stage interaction (depicted in Figure 36.1). The main dining room is the front-stage, where the server (social actor) is performing a role that includes cultural norms of polite language and respectful service to the patrons (audience). The back-stage is the kitchen, where the server is not expected to perform norms of politeness and can be more informal, even expressing frustration or stress while no longer presenting the ideal server role to the audience of patrons.

Front-stage interaction in a restaurant dining room

"Actor" presents idealized self-performance with a focus on appearance, manners, and social status. Interaction is based on role expectations and evaluation by an "audience."

Back-stage interaction in a restaurant kitchen

"Actor" can relax without concern for idealized self-performance (ie, show frustration) because interaction is informal and free from evaluation or judgement by an "audience."

Figure 36.1 Examples of front-stage and back-stage interaction in a restaurant.

36.1.02 Impression Management

Impression management involves all the conscious and unconscious ways that individuals try to present themselves to others, including appearance, speech, and behavior. Individuals carefully manage the presentation of self to project a positive self-image in interaction. For example, when individuals go on first dates, they are intentional about what they wear, how they speak, and what aspects of themself they share with their romantic interests.

The social situation directs which impression one wants to give to others in interaction. Different roles and scripts are required in different social environments, and a particular self-performance (eg, cheering rock music fan) may be positive in one setting (eg, at a rock concert) but not in another (eg, at a business meeting). The importance of social context directly connects to the concept of dramaturgy (see Figure 36.2). Understanding the "stage" helps a social actor know how to perform in a desirable way and to manage the evaluation of the presentation of self.

Formal setting (eg, business meeting)

Individual presents themself in a professional way through gestures, speech, and dress.

Less formal setting (eg, concert)

Individual presents themself in informal, fun, and energetic way through gestures, facial expressions, and attire.

Figure 36.2 Examples of impression management based on social setting.

Lesson 36.2
Status

36.2.01 Status and Status Set

Status refers to the social position an individual holds in society. Statuses are a part of social systems (eg, family, workplace, healthcare) and exist in relation to one another (eg, parent/child, employee/manager, physician/patient). In addition, many statuses are organized in a hierarchical structure in which some statuses (eg, manager) have more power than others (eg, employee).

Every individual has multiple statuses, which are, together, referred to as a **status set**. The social setting, rather than the individual, determines which status is present and necessary for interaction. For example, an individual's status set could include roommate, student, and sibling, and the social setting determines whether the individual is in the status of roommate, student, or sibling. Each status is associated with specific expectations for how one is to behave and interact with others. These expectations are further discussed in Lesson 36.3.

36.2.02 Types of Status

Statuses can be classified as ascribed, achieved, or master (see Figure 36.3). An **ascribed status** is an involuntary social position assigned by society that is typically based on social identity categories such as sex, race, or nationality. Ascribed statuses are often lifelong social positions.

Alternately, an **achieved status** is a *voluntary* social position earned through merit or choice and reflects accomplishment, skill, and/or ability. Examples of achieved statuses include merit-based positions such as college graduate and physician, as well as chosen social positions such as friend and volunteer.

A **master status** overshadows the other statuses in an individual's status set. Other people tend to view an individual as their master status in all interactions, even in situations that would otherwise be guided by another status. A master status can be either ascribed or achieved, and an individual may have more than one master status. For example, an individual's occupational status of physician may be a master status when other people, such as a neighbor, view her as a physician outside of the workplace. However, if others do not know that she is a physician, her sex (ie, female) may become the master status in an interaction.

Figure 36.3 Examples of types of status.

Lesson 36.3
Roles

36.3.01 Roles

Status refers to the social position an individual holds in society (see Lesson 36.2). Every status has an associated **role** (also known as social role) that describes the expected behaviors of individuals in the status. For example, "student" is a status, and the associated role would include expected behaviors such as attending class, completing assignments, and studying for exams.

A **role performance** occurs when an individual acts out the expected behaviors of a role in social interaction. At times, individuals may not complete all expected behaviors (eg, a professor does not finish grading exams on time), resulting in a less successful role performance. Because individuals have multiple roles, attempting to perform the numerous behaviors expected of them can produce situations of role strain and role conflict (as detailed in Concept 36.3.02).

Roles are a part of social structure, meaning that they are tied to the social situation rather than to the individual performing the role. In addition, the expected behaviors of a role remain consistent in all social situations no matter who is performing the role. For instance, in the U.S. government, the "President" is a role with various expectations for behavior (eg, sign/veto bills, address national issues). The set of expected behaviors (see Figure 36.4) remains consistent and applies to every individual performing the role of President, regardless of the individual characteristics of those who hold the office.

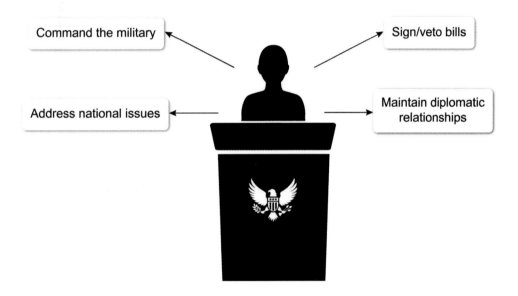

Figure 36.4 Examples of role expectations for the President of the United States.

36.3.02 Role Strain, Role Conflict, and Role Exit

Role strain occurs when the expectations for a *single role* compete, producing tension. Role strain describes a situation in which an individual is overloaded with expectations for a single role, so performing all expected behaviors is difficult. For example, role strain could occur for a mother (ie, role) who struggles to balance the expected behaviors of driving her children to after-school activities and cooking nutritious meals.

On the other hand, **role conflict** occurs when the expectations for *two or more roles* simultaneously held by an individual compete, producing tension. Role conflict creates a situation in which an individual is faced with opposing expectations from each role and must choose which role to perform. For example, role conflict could occur for a physician who is also a husband and father. The expectation for his role as physician to work on weekends may conflict with expectations for his role as parent, such as attending his child's soccer game, and his role as husband, such as spending time with his spouse.

Role strain and role conflict describe two different types of role challenges individuals experience due to tension created by competing expectations for behavior (see Table 36.1). One response to this tension is **role exit**, which occurs when an individual moves from one role to another, disengaging from the original role's expected behaviors and taking on the expected behaviors of a new role. For example, a physician who retires from a long career at a busy hospital to travel is experiencing role exit by disengaging from her occupational role and starting a new role as a retiree.

Table 36.1 Types of role challenges.

Role strain	<ul><li>Competing expectations within a single role create tension<ul><li>Example: a student struggles to find enough time to complete all readings before class (expectation 1) and write lab reports (expectation 2)</li></ul></li></ul>
Role conflict	<ul><li>Competing expectations for two or more roles create tension<ul><li>Example: a student (role 1) employed part-time (role 2) struggles to find enough time to complete homework and work late hours, respectively</li></ul></li></ul>
Role exit	<ul><li>Individual disengages from a role, often replacing it with a new role with new expectations for behavior<ul><li>Example: a college student (old role) graduates and begins full-time employment (new role)</li></ul></li></ul>

Lesson 36.4
Groups

36.4.01 Primary and Secondary Groups

A **social group** (also called a group) is a set of individuals (ie, two or more people) who engage in interactions based on shared experiences or goals. Groups are an important element of society because most social interaction occurs within group settings.

Primary groups are composed of a small number of individuals who share close, personal relationships that typically involve face-to-face interactions. In primary groups, members are united by emotional bonds and typically have long-lasting, committed relationships. Examples of primary groups include families, close friends, and couples in romantic relationships.

On the other hand, **secondary groups** are composed of individuals who share impersonal relationships that involve interactions aimed at accomplishing a task (see Figure 36.5). In secondary groups, members are united by shared goals and typically disband after the group's goals are achieved. Examples of secondary groups include exam study groups and professional associations.

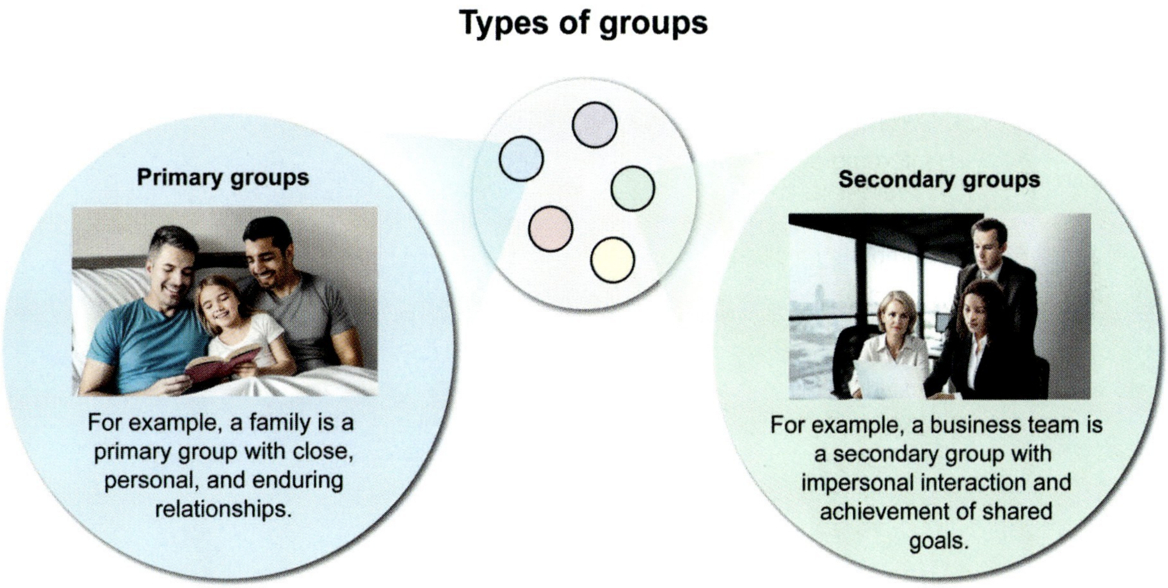

Types of groups

Primary groups

For example, a family is a primary group with close, personal, and enduring relationships.

Secondary groups

For example, a business team is a secondary group with impersonal interaction and achievement of shared goals.

Figure 36.5 Examples of primary and secondary groups.

36.4.02 In-Group versus Out-Group

Groups can also be categorized based on membership. An **in-group** is a group to which individuals feel a sense of belonging, whereas an **out-group** is a group to which individuals do not feel they belong. In-groups and out-groups can be defined by social identity categories (eg, race, class) or other criteria such as shared interests (eg, sports team fans) or geographic region (eg, hometown).

Individuals tend to view their own in-group favorably and engage in comfortable interactions with these members. Alternatively, out-groups tend to be viewed unfavorably, and, at times, interaction with out-groups is based on antagonistic feelings (eg, opposing political parties).

Individuals often compare themselves to others. A **reference group** refers to comparison groups used to evaluate one's own behavior and self-concept. For example, an individual who compares their medical school entrance exam study habits to those of other pre-med students is using the other students as a reference group. Individuals may be members of the reference group (eg, family, friends), or an individual may aspire to emulate members of the reference group (eg, celebrities, leaders). Individuals typically have multiple reference groups at a given time, and these also change throughout life.

36.4.03 Group Size

Another difference between groups is size (ie, number of individual members), which impacts the strength and closeness of the group. The relationships between individual group members are called social ties. As **group size** increases, the potential number of social ties increases as well. Larger groups (eg, whole societies) are stronger due to the higher number of social ties; however, in smaller groups (eg, roommates) members are more reliant on one another and have closer relationships.

The smallest group is a **dyad**, which is composed of two individuals with only one social tie. Dyads are the simplest and most intimate group because the two members depend solely on each other. However, dyads are less stable than larger groups because if either individual leaves, the group ceases to exist. For example, a husband and wife are a dyad with one, intimate social tie to each other, and if either individual chooses to leave the group (ie, divorce), the dyad dissolves.

A **triad**, which is a group of three individuals, is less intimate than a dyad because there are three social ties, and the members can depend on two other individuals. As such, triads are more stable than dyads because the greater number of social ties allows one individual to leave the group, but the group would still exist (as a dyad). For example, if a husband and wife have a child, the group becomes a triad with three social ties and is more stable because one individual can leave, but the group would continue to exist. Figure 36.6 shows the key differences in groups based on size.

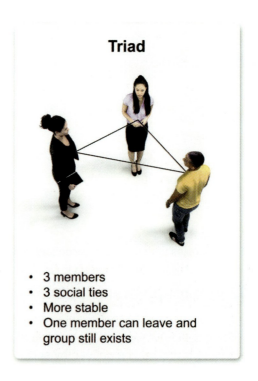

Figure 36.6 Comparing features of dyad versus triad groups.

Lesson 36.5
Networks

36.5.01 Networks and Social Network Analysis

A **social network** is an informal and nonhierarchical web of connections between individuals, groups, and/or organizations. Networks consist of an assortment of direct (eg, spouse, neighbor) and indirect (eg, friend of a friend) relationships linking individuals to one another. **Strong ties** refer to solid connections between an individual and family or close friends, whereas **weak ties** are loose connections between acquaintances or coworkers. Figure 36.7 is an example of a social network with multiple nodes and ties.

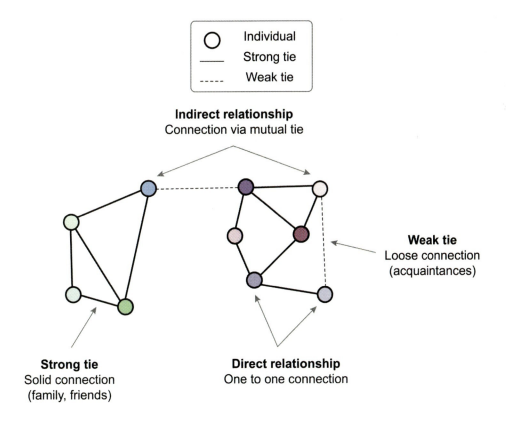

Figure 36.7 Social network example.

Researchers use these visualizations in **social network analysis**, which examines the organization and function of social networks by mapping the patterns of interaction between network ties. For example, epidemiologists may use social network analysis to examine the transmission of infectious diseases, such as contact tracing during the coronavirus pandemic. Epidemiologists analyze network ties to help illustrate the possible trajectory of the illness and recommend isolation for infected and exposed individuals to help reduce further contagion.

36.5.02 Social Capital and the Strength of Weak Ties

The connections individuals have in their networks can help them advance in society. The concept of **social capital** (see Concept 45.1.03 for more detail) describes the potential value embedded in an individual's network connections. Social capital can confer advantages, depending on who is part of the network and what resources (eg, information, opportunities) those network ties provide to help the individual. For example, if an individual has an aunt whose college roommate is the president of a prestigious medical school, the network connection could provide access to information to help them get admitted into the medical school.

In addition, weak ties can connect an individual to social capital embedded in networks with which they have no direct relationships. The term "the strength of weak ties" describes the value of loose connections and indirect relationships. Strong ties can provide social support but often have overlapping network connections that do not provide additional resources for an individual. However, weak ties can offer access to new networks that may provide valuable opportunities, especially for employment, such as having a friend of a friend who works at the hospital where one hopes to get a job (shown in Figure 36.8).

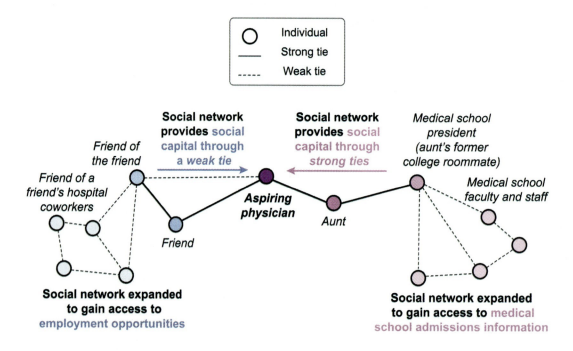

Figure 36.8 Social capital and the strength of weak ties example.

Lesson 36.6
Organizations

36.6.01 Types of Organizations

In sociology, **organization** is a broad term used to describe various forms of complex groups (typically large secondary groups, see Concept 36.4.01) that cooperatively interact to achieve a specific purpose. Much of contemporary society depends on the work of organizations, and individuals' lives frequently include experiences within organizations (eg, workplaces, stores, community groups).

Organizations are designed in different ways and serve different purposes (see Figure 36.9). **Informal organizations** have a loose structure without clearly defined roles or rules. This type of organization is typically based on shared interests or needs and involves regular interaction but may be temporary. For example, a neighborhood children's playgroup is an informal organization convened to arrange play in a community but does not include specific roles or rules.

Most organizations in contemporary society are **formal organizations** with explicit policies and roles in place to effectively achieve the organization's goals. This type of organization is designed to provide stability for its members and promotes reliable ways to complete activities. For example, a hospital is a formal organization designed to systematically treat patients using different roles (such as nurse and doctor) and clear procedures for effective care.

Informal organization
(eg, neighborhood playgroup)

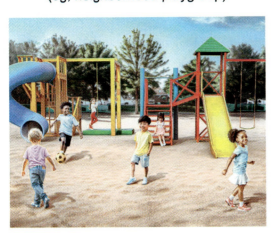

Based on shared interests, with no formal rules or roles

Formal organization
(eg, hospital system)

Based on explicit policies and roles to achieve the organization's goals

Figure 36.9 Examples of informal and formal organizations.

36.6.02 Bureaucracy

A **bureaucracy** is a specific type of formal organization designed to maximize efficiency and productivity through standard policies and specialized tasks (eg, government agencies, large corporations). Sociologist Max Weber argued that bureaucracies are needed to manage the complex social systems of modern societies. Weber researched the common features of bureaucracies and developed an **ideal**

type (ie, set of essential characteristics). Actual bureaucracies often have some, but not necessarily all, of the characteristics of the ideal type, which include the following:

- **Division of labor** increases efficiency through specialization in which individuals focus on a particular area of expertise.
- **Hierarchical structure** provides a clear chain of command, enabling everyone to understand their role.
- **Standardization** of clearly written rules and regulations ensures predictability and uniformity.
- **Impersonal** and **impartial** procedures for hiring/promoting/firing are based on merit.
- **Written records** of all operations are kept to monitor productivity and effectiveness.

For example, a school system is a bureaucracy with a division of labor including teachers, administrators, and cafeteria staff with specialized tasks, and the roles are organized in a hierarchy with administrators leading the bureaucracy. There are clear rules and processes outlined in employee manuals, and administrators strive for impartiality in hiring practices (eg, hiring the most qualified candidate). Finally, written records of student grades and faculty performance are maintained to evaluate effective teaching and achievement of learning outcomes.

Although bureaucracies are carefully planned organizations, some have been critiqued as inflexible and difficult to navigate due to the rigid rules and procedures (eg, the multiple steps required to register a car in a new state). Another critique, called the "**iron law of oligarchy**," describes problems in bureaucratic leadership. Oligarchy refers to "ruling by the few," and, over time, power within a bureaucracy becomes concentrated among a small number of leaders when the desire for personal power surpasses the organization's shared goals.

36.6.03 McDonaldization

The characteristics of bureaucracies were useful in the development of early industrial systems, such as the division of labor on the automobile assembly line. More recently, other industries, including fast-food restaurants, have adopted similar processes to ensure that all franchises produce standardized products and experiences. Sociologist George Ritzer developed the concept of **McDonaldization** to describe the trend to incorporate elements of bureaucracies in many parts of society.

McDonaldization includes four common features aimed to produce extreme productivity and reliability within organizations. However, unintended negative consequences are associated with each of the features. **Efficiency** produces optimization of the quickest and least expensive systems needed for the operations of the organization (eg, assembly-line production). Designing efficient systems comes at the cost of individuality in which personal or individual needs are not accommodated (eg, personalizing an order is not available).

McDonaldized organizations focus on **calculability**, which is used to understand the operations of the organization via measurable units, ensuring high quantities of production. For example, a fast-food restaurant has standard metrics for the time required to prepare each item, and each customer's order is timed to calculate the effectiveness of the system. The consequence of a focus on high quantities is a decrease in the quality of the product and experience.

The feature of **predictability** ensures that all systems are uniform, which produces standardization of customer experiences. For example, the interactions at a fast-food restaurant are consistent and as expected because employees follow a script when taking orders and the store layout is the same at all locations. However, standardization of products and customer experiences can reduce the potential for uniqueness and/or innovation in the systems of operation.

Last, **control** increases automation through reliance on non-human tools and technology such as using machines to cook or phone applications to order. Although an increased reliance on automation provides greater control of processes, the use of technology also reduces the need for a skilled workforce.

Table 36.2 outlines the four main features and consequences of McDonaldization within fast-food restaurants, as well as the practice of medicine.

Table 36.2 Features of McDonaldization.

	McDonaldization			
	Intended result	**Unintended result**	**Fast-food example**	**Medicine example**
Efficiency	↑ Optimization	↓ Individuality	Customer orders at register and cleans own table	Patient goes to a "Minute Clinic" for specific concerns
Calculability	↑ Quantity	↓ Quality	Store success measured by number of products served	Physicians compensated based on number of patients seen
Predictability	↑ Uniformity ↑ Standardization	↓ Uniqueness	All restaurant chains appear the same, have same products	Standardized patient checklists make appointments uniform
Control	↑ Automation	↓ Skilled workforce	Automated machines cook products	Electronic patient portals allow patients to see test results

Lesson 37.1
Attraction

37.1.01 Attraction

Interpersonal attraction, which is defined as liking, or positive feelings toward another person, impacts relationships of all kinds (eg, professional, friendship, romantic).

Studies have shown that multiple factors increase attraction (ie, liking) toward another person. These factors include: similarity (how similar the other person is to oneself, such as in attitudes or personality), physical attractiveness, believing one is liked in return, and/or proximity. Proximity (Figure 37.1) refers to being geographically close (eg, neighbors, classmates); mere exposure (being around the other person often) contributes to proximity.

For example, two girls went to the same middle school and summer camp.

By the time they got to high school, the girls had become friends, demonstrating the role of proximity in interpersonal attraction.

Figure 37.1 Example of the role of proximity in interpersonal attraction.

Lesson 37.2
Aggression

37.2.01 Aggression

In social psychology, **aggression** is defined as behavior intended to harm another. One brain area that is important in aggression is the amygdala (Concept 4.3.01). Researchers have shown that electrical stimulation of the amygdala can lead to displays of aggression.

The **frustration-aggression theory** contends that individuals exhibit aggressive behaviors as a result of having a goal or effort blocked or defeated (ie, frustration). For example, a child who is not allowed to play with a desirable toy displays violent behavior toward a peer.

Additionally, as Concept 18.1.01 introduces, the "Bobo doll" experiments demonstrated that social learning contributes to aggression. These studies showed that children who observed others acting aggressively toward a "Bobo doll" were more likely to display aggression towards the doll themselves.

Lesson 37.3
Attachment

37.3.01 Attachment

Attachment reflects the emotional bond between child and caregiver. **Securely attached** children confidently explore their environment and return to their caregivers as a consistent and nurturing base. If their caregivers leave, securely attached children become upset but calm quickly. When their caregivers return, securely attached children express pleasure and seek contact.

In contrast, **insecurely attached** children act indifferent or clingy toward caregivers and may show limited exploration. If their caregivers leave, insecurely attached children may act indifferent or become extremely upset and difficult to calm. When their caregivers return, insecurely attached children may act clingy or avoid contact. Secure and insecure attachment styles are compared in Table 37.1.

Table 37.1 Secure versus insecure attachment styles.

Attachment style	Behaviors exhibited
Securely attached children	• View their caregivers as a consistent and nurturing base • Confidently explore their environment
Insecurely attached children	• Act indifferent or clingy toward caregivers • May show limited exploration of their environment

Mary Ainsworth, a developmental psychologist, studied these attachment styles through the "Strange Situation" experiment. In this study, infants were left alone with a stranger and then reunited with a caregiver in order to reveal the infants' attachment styles.

Additionally, in a series of studies on the role of contact comfort in attachment, Harry Harlow examined the attachment of baby monkeys to artificial mothers. He found that baby monkeys spent more time with an artificial mother made of cloth that did not provide food than with one made of wire that did provide food.

Lesson 38.1

Altruism

38.1.01 Altruism

Altruism (Figure 38.1) is defined as unselfish concern or behavior intended to benefit others. Altruistic behavior may even come at the expense of one's own well-being or safety. For example, a vampire bat that has been able to feed recently will sometimes share blood with an unsuccessful bat that is starving, leaving less blood for itself (an unselfish behavior that decreases its own food supply).

For example, altruism is demonstrated when a satiated vampire bat unselfishly shares blood with a starving vampire bat, leaving less blood for itself.

Figure 38.1 Altruism example.

Lesson 39.1

Attributional Processes

39.1.01 Attributional Processes

Attribution theory states that individuals assign reasons (ie, attributions) for behavior. Often, these attributions are either internal (ie, dispositional) or external (ie, situational). For example, one could blame getting cut off in traffic on internal factors (eg, "that driver is a jerk") or external factors (eg, "that driver couldn't see me because of the sun") (Figure 39.1).

Attributions can be internal (ie, dispositional) or external (ie, situational).

For example, a truck driver cuts off a cyclist (behavior) and the cyclist attributes the truck driver's behavior to...

...an internal factor (ie, being a jerk).   ...an external factor (ie, the sun).

Dispositional attribution

Situational attribution

Figure 39.1 Situational versus dispositional attributions.

Attributional biases are common cognitive biases that may occur when individuals assign attributions. For example, **self-serving bias** occurs when one's own success is attributed to internal factors (eg, winning as a result of talent), whereas one's own failure is attributed to external factors (eg, losing because of unfair refereeing) (Figure 39.2).

Success attributed to internal factors

Failures attributed to external factors

Figure 39.2 Example of self-serving bias.

The **fundamental attribution error** occurs when an individual assumes that someone else's behavior is the result of internal (ie, dispositional) rather than external (ie, situational) factors. For example, a soccer player decides that another player's behavior (eg, committing a foul) was caused by temperament (eg, rudeness), an internal factor (Figure 39.3).

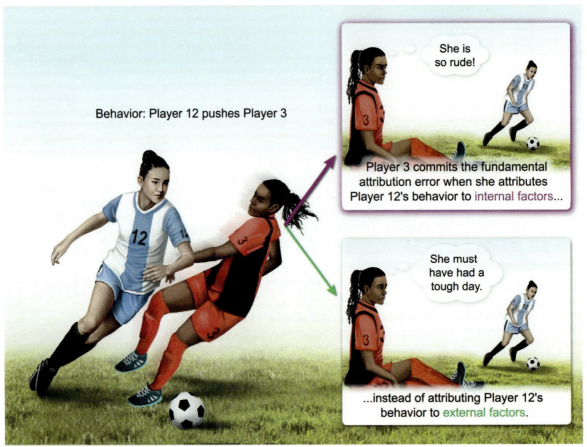

Figure 39.3 Fundamental attribution error example.

An attribution error called the halo effect occurs when an individual attributes additional positive qualities (eg, intelligence) to a person with one positive quality (eg, physical attractiveness). For example, when research participants were shown photographs of attractive and unattractive individuals and asked to choose which person was more intelligent, they tended to judge the attractive individuals as more intelligent than the unattractive individuals.

Another attributional bias, **actor-observer bias** (Figure 39.4), occurs when an individual attributes another person's behavior (eg, cutting someone off in traffic) to internal factors (eg, "that person is a jerk") while attributing one's own behavior to external factors (eg, "I am running late").

For example, when an individual is cut off in traffic by another driver (ie, another person's behavior)...

However, when the individual cuts off another driver in traffic (ie, own behavior)...

...the individual assumes the other driver is a jerk (ie, internal factor).

...the individual attributes his own behavior to running late (ie, external factor).

Figure 39.4 Example of actor-observer bias.

Additionally, the **just-world hypothesis** describes a tendency to assume that bad things happen to people who deserve them. For example, an individual who finds out that her friend was diagnosed with a serious illness may assume that her friend did something wrong to deserve the illness (ie, an attributional bias). By perceiving the victim (her friend) as deserving of the misfortune (a serious illness), the individual maintains her belief in a "just" (fair) world.

Lesson 40.1
Prejudice

40.1.01 The Contributions of Power, Prestige, and Class in Prejudice

Prejudice is a preconceived, negative belief or feeling about individuals or groups based on generalizations about their group membership. These beliefs are often not based on experience or evidence but are learned through interactions with others.

When individuals are socialized into groups, they learn about differences in behaviors and characteristics between in-groups and out-groups (see Lesson 36.4). Prejudice can be learned by defining groups based on an "us" versus "them" approach. For example, prejudice can develop between groups who support opposing political candidates, and each group may hold negative attitudes toward one another.

In addition, social inequalities (described in Lesson 45.1) and the hierarchical organization of individuals in society based on power, prestige, and class can contribute to prejudice:

- **Power** is the ability to act based on one's own interests to achieve goals without restrictions. Certain careers (eg, politician) and accomplishments (eg, a large social media following) increase one's power.
- **Prestige** refers to the amount of respect one receives based on social position. Certain occupations (eg, physician), personal characteristics (eg, attractiveness), and achievements (eg, winning an Olympic gold medal) confer prestige.
- **Class** is largely determined by economic resources (eg, income, property). Wealthy individuals are at the top of the social hierarchy, whereas those in the working and lower classes are at the bottom.

Individuals and groups with less power and prestige often experience prejudice from more powerful others. For example, the belief that all individuals who live below the poverty line are "uneducated" and "lazy" is a prejudicial attitude held by some individuals with a higher class position.

40.1.02 The Role of Emotion in Prejudice

An individual's emotions (see Chapter 24) also play a role in the development of prejudice. When faced with negative emotions such as fear or frustration, individuals often feel stronger connections to their in-group, which can lead to increased feelings of prejudice toward out-groups. For example, during an international war, individuals may develop fear or anger toward the adversary nation, and those emotions can contribute to prejudicial feelings about those who live in that nation (ie, an out-group).

40.1.03 The Role of Cognition in Prejudice

Cognition (see Chapter 20) involves higher-order mental processes, including attention, memory, language, thinking, and problem-solving. Some types of thinking and cognitive processes can contribute to prejudice.

The brain quickly categorizes things (including people) using **schemas**, which are mental frameworks that organize old information and allow quick processing of new information. This cognitive process can lead to prejudice because people tend to unconsciously and automatically categorize others based on their most obvious social identity categories, including age, race/ethnicity, and gender. For example, when seeing a white-haired man walking with a cane, an individual may categorize the man as "old" and assume that he is also "weak" and "frail."

Lesson 40.2
Stereotypes

40.2.01 Stereotype Threat

Similar to prejudice (Concept 40.1.01), **stereotypes** are generalized beliefs about groups of people; but unlike prejudice, which is always negative, stereotypes can be positive, negative, or neutral. Stereotypes often develop based on limited interactions and/or simplified information and are applied to all individuals within the social group. For example, expecting all people who live in rural areas to be farmers is a stereotype.

Stereotype threat refers to the anxiety experienced by an individual who feels judged based on a negative stereotype about a group to which they belong. For stereotype threat to arise, the individual must be made aware of their membership in a group that is negatively stereotyped. This awareness arouses vigilance and often results in anxiety that negatively impacts performance. For example, female students perform more poorly on math tests when they are first reminded of the stereotype "girls are bad at math."

40.2.02 Self-Fulfilling Prophecy

When an individual belongs to a group that is stereotyped, they may internalize the stereotype. A **self-fulfilling prophecy** occurs when a belief (which may or may not be true) influences behavior such that the belief becomes true. For example, an art major believes he is terrible at math, so he does not study very hard for the math final because he thinks, "What's the point in trying so hard? I'm terrible at math and will fail this test no matter what I do." Then, his lack of preparation causes him to fail the exam, thus supporting his belief that he is bad at math.

The self-fulfilling prophecy phenomenon can be applied to beliefs individuals hold about other people or groups. For example, a teacher believes a student will not perform well in class and therefore pays less attention to the student, who then does poorly in class.

40.2.03 Stigma

A **stigma** is any characteristic (eg, laziness, bodily disfigurement) that is devalued and/or labeled as abnormal or unacceptable in society. Stigma can affect an individual's identity (eg, internalization of negative label) as well as how they are treated by others (eg, isolation, mistreatment).

For example, some medical conditions, including obesity and lung cancer, are stigmatized because of their association with socially disliked behaviors (ie, overeating and smoking, respectively). As a result, patients with these conditions may experience mistreatment in medical interactions (eg, clinicians may focus on the patient's body weight while overlooking other aspects of the patient's health).

40.2.04 Ethnocentrism and Cultural Relativism

At times, cultural beliefs and practices can be used as the foundation for stereotypes. **Ethnocentrism** is the belief in the superiority of one's own culture (Lesson 34.1), which results in evaluating other cultures based on one's own cultural values and practices. For example, an individual judging a cultural group who eats rodents as "strange" or "gross" because the food differs from their own culture is engaging in ethnocentrism.

The opposite of ethnocentrism, **cultural relativism**, suggests that there are no "right" or "wrong" cultural practices. Cultural relativism advocates examining a culture based on its own context rather than comparing it to another culture. For example, Muslim women wearing the traditional veil (ie, hijab) is a cultural practice. Judging this practice as "oppressive" compared to mainstream U.S. culture is ethnocentric but viewing the practice within the larger context of Muslim culture, without judgment, is an example of cultural relativism.

These two approaches to cultural differences have ramifications outside of individual experiences. Ethnocentrism can produce ideas of cultural superiority that can lead to acts of discrimination (see Lesson 40.3) or conflict. On the other hand, cultural relativism can encourage inclusive practices and policies to foster positive interactions between cultural groups.

Lesson 40.3
Discrimination

40.3.01 Prejudice and Discrimination

Discrimination is the unjust treatment of individuals based on their membership in a social group; those who are discriminated against are not given equal access to resources and opportunities. Prejudice (ie, negative beliefs or attitudes about an individual based on group membership as defined in Lesson 40.1) and discrimination are related concepts. For example, an individual may hold prejudice toward women such as the belief that "women aren't good leaders." If that individual were to take action based on this prejudice, such as not giving a leadership position to a woman, discrimination has occurred. However, prejudice does not always result in discrimination.

Concept 40.1.01 describes how prejudice is formed based on inequalities related to power, prestige, and class. Similarly, acts of discrimination are also shaped by inequalities wherein those with less power and lower social standing (ie, prestige) experience discrimination and thus lack the resources to advance within the social hierarchy.

40.3.02 Individual versus Institutional Discrimination

Discrimination can take place at the individual level (ie, one person acts based on prejudice) and at the institutional level (ie, institutional policies/procedures are shaped by prejudice). **Institutional discrimination** is the unjust treatment of specific social groups (eg, racial/ethnic minorities) built into the framework of organizations and institutions rather than held as individual beliefs and attitudes. For example, institutional discrimination occurs when police department policies (rather than individual officers' decisions) cause resources, such as patrol officers, to be focused on predominantly Black neighborhoods.

Institutional discrimination can take place within any institution (eg, schools, hospitals, corporations) and includes actions and policies that are often subtle and/or unintentional but nevertheless cause harm. For example, if employees can only be promoted if they do not take extended time off from work, those who take maternity leave are subtly discriminated against because they are effectively prevented from promotion based on this policy.

Lesson 41.1

Group Processes

41.1.01 Social Facilitation

Human behavior can be strongly influenced by social factors, such as the presence of others. **Social facilitation** is defined as an improvement in performance of well-rehearsed or easy tasks in front of a crowd versus when alone. For example, a basketball player who frequently practices free throws in an empty gym may shoot free throws more accurately when there are other people in the gym (Figure 41.1).

A basketball player frequently practices shooting free throws alone...

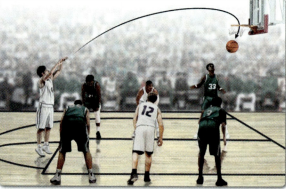
...but he shoots free throws more accurately when there are other people in the gym.

Figure 41.1 Social facilitation example.

41.1.02 Social Control

Social control is the exertion of power by a group, institution, or society to ensure that the behavior of individuals conforms to particular norms (ie, expected behaviors). For example, the physical layout of a grocery store, which is designed so that shoppers will walk in a certain direction and see items in a particular order, exerts social control.

Lesson 41.2

Social Loafing, the Bystander Effect, and Deindividuation

41.2.01 Social Loafing

Social loafing occurs when individuals exhibit less effort on a task when part of a group than when alone. For example, while working on a group presentation, a group of coworkers puts in less effort than when they prepare presentations individually (Figure 41.2).

For example, while working on a group presentation, coworkers put in less effort...

...than when they prepare presentations individually.

Figure 41.2 Social loafing example.

41.2.02 The Bystander Effect

The **bystander effect** occurs when people are less likely to help someone in need if other people are around. This phenomenon is related to the *diffusion of responsibility*, which is when individuals assume less responsibility for taking action when they are around others. For example, several people may witness a woman falling in a crowded gym and not stop to help her (Figure 41.3).

| For example, while running on a treadmill in a crowded gym, a woman trips and falls. | Although there are many people nearby, no one stops to help, demonstrating the bystander effect. |

Figure 41.3 Example of the bystander effect.

41.2.03 Deindividuation

Being part of a large, animated group can increase an individual's arousal and lead to **deindividuation**, which is the loss of self-awareness, inhibition, and sense of personal responsibility. For example, a spectator in a large crowd at a football game, feeling energized and with less personal awareness and restraint, may say something he normally would not.

Lesson 41.3

Conformity and Obedience

41.3.01 Conformity

Conformity is the adjustment of one's behavior or thinking to align with that of a group. An individual may conform in an attempt to fit in (ie, belong) or to avoid rejection. For example, a person who dislikes a particular food might eat that food at a party out of a desire to fit in with others who are eating it.

Solomon Asch conducted experiments on conformity (Figure 41.4); in Asch's studies, participants were asked to pick which comparison line was the same length as a standard line. When alone, participants almost always selected the correct line. However, when with several confederates (ie, actors) who all selected an incorrect line, participants also picked the clearly wrong line over 30% of the time.

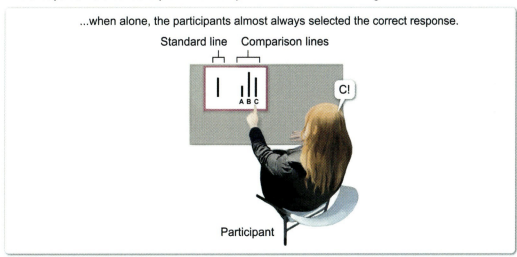

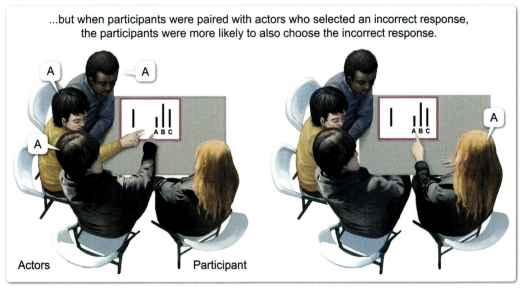

Figure 41.4 Solomon Asch's conformity experiment.

41.3.02 Obedience

Compliance occurs when one individual modifies their behavior at the request of another individual. For example, a student switches seats when asked to by a classmate. **Obedience** is a type of compliance that occurs when an individual carries out behavior based on the orders of an authority figure (eg, mopping the floor when one's manager says to do so).

In Stanley Milgram's experiments on obedience, the participant was instructed by the researcher (ie, an authority figure) to administer a shock to a learner (an actor) each time they answered a question incorrectly. The shocks increased in intensity with each incorrect response. Milgram found that most people (about 60%) delivered the highest intensity shock to the learner if commanded to do so, even though the final shocks were marked "Danger: Severe Shock" and "XXX."

Lesson 41.4

Group Decision-Making

41.4.01 Group Polarization

Group polarization occurs when the average attitude or opinion of group members becomes more extreme after group discussion (Figure 41.5). Group polarization is more likely to occur when group members share a similar opinion at the beginning of the discussion. The group opinion may become polarized in either direction (ie, more positive or more negative). For example, a group of donors who support a charitable organization may feel even more favorably about the organization after discussing it with fellow supporters.

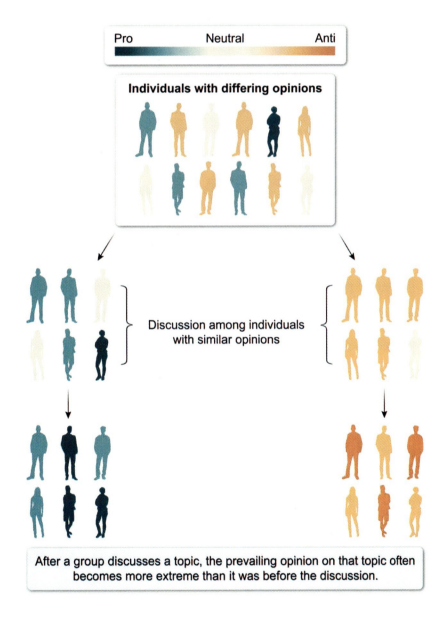

Figure 41.5 Group polarization.

41.4.01 Groupthink

Groupthink occurs when a desire to maintain group cohesion (ie, bonds among group members) and reach a consensus outweighs critical decision-making. For example, in a class, several friends are assigned to a group and instructed to choose a position on an issue; the friends have differing opinions on which side of the issue is best but choose one position without much discussion because they do not want to disagree with each other (Figure 41.6).

For example, group members have differing opinions about which side of an issue is best but choose one position without much discussion for the sake of the group.

Figure 41.6 Example of groupthink.

Lesson 42.1
Social Norms

42.1.01 Folkways, Mores, and Taboos

Concept 34.4.01 describes socialization as the lifelong process of learning the expected behaviors (ie, norms) and beliefs of one's own society. In sociology, **norms** are unwritten rules for behavior that people in society are expected to follow (eg, chewing with a closed mouth). When individuals follow social norms, societal order is established by ensuring that behavior is predictable (shown in Figure 42.1). Norms also help individuals decode and understand the behavior of others (eg, the norm of silence in a library helps explain why a friend does not start a conversation there).

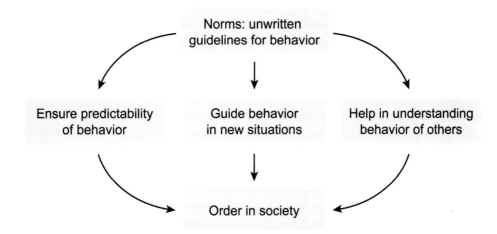

Figure 42.1 Impact of norms on guiding behavior and establishing order in society.

Cultural values (see Lesson 34.1) guide the development of norms because expectations for behavior reflect what is important and moral in society. Norms exist in every culture but can vary based on differences in cultural values and beliefs. For example, in the United States, there is a norm to work for eight hours a day with brief breaks. However, in some Latin and Mediterranean cultures, there is a norm to take a long afternoon break during the workday for rest and leisure.

Four different types of norms reflect differences in how important the expected behavior is within a culture:

- **Folkways** are informal normative behaviors and/or traditions in society (eg, face forward on an elevator).
- **Mores** are more serious norms with greater moral significance and a stronger connection to cultural values (eg, original work on homework assignments).
- **Laws** are formal norms strictly enforcing what society views as right and wrong (eg, must be 21 years old to drink alcohol in the United States).
- **Taboos** are society's strongest norms (eg, do not eat another human), and violations are considered morally reprehensible.

42.1.02 Sanctions

Associated with norms, **sanctions** promote compliance (Concept 41.3.02) with normative behaviors and are a means of social control (Concept 41.1.02) in society. A **positive sanction** is a reward for upholding a norm, whereas a **negative sanction** is a repercussion for violating a norm. (Note: sociology uses the terms "positive" and "negative" differently than how psychology uses these terms; for example, see Concept 17.2.01.)

For example, parents may establish a norm of quiet behavior for their child in public settings. Giving the child a piece of candy for quiet behavior in the grocery store would be a positive sanction (ie, a reward for complying with the norm), whereas scolding the child for yelling in a grocery store would be a negative sanction (ie, repercussion for violating the norm).

In addition, sanctions can be enforced through informal and formal mechanisms. **Formal sanctions** are codified within the institutions of society, such as policies or laws (eg, repeated absences from work result in termination). **Informal sanctions** are not codified and are enforced by members of social groups (eg, marital infidelity results in social shunning from the community).

All four possible variations of sanctions (ie, positive formal, positive informal, negative formal, negative informal) serve to promote socially expected behaviors (shown in Table 42.1).

Table 42.1 Examples of the four types of sanctions.

	Formal Sanctions (codified in institutions)	**Informal Sanctions** (enforced in social groups)
Positive Sanctions (rewards)	**Positive formal** Institutional reward for upholding a norm (eg, city council presents a citizenship award for volunteering at a hospital)	**Positive informal** Social reward for upholding a norm (eg, friend gives a high five for telling a funny joke)
Negative Sanctions (repercussions)	**Negative formal** Institutional repercussion for violating a norm (eg, police officer issues a ticket for speeding)	**Negative informal** Social repercussion for violating a norm (eg, teacher scolds a student for talking during class)

42.1.03 Anomie

At times, there are situations where norms may no longer have significance or the ability to regulate individual behavior in a society. Sociologist Émile Durkheim developed the concept **anomie** to describe this state of normlessness. When the norms and values of a society are challenged but have yet to be replaced, the social system reaches the state of anomie. Without norms, individuals lack guidance and purpose and may develop feelings of aimlessness or disconnection from society.

Anomie is often the result of a shift or transition in society that causes instability. For example, during the Industrial Revolution, many individuals left their lives in agricultural communities and moved to cities to start a new way of life working in factories. This societal shift resulted in a period of anomie because the norms governing life prior to the Industrial Revolution (eg, contribute to community work, live with extended family) were no longer effective or relevant.

Lesson 42.2

Deviance

42.2.01 Deviance

While norms are necessary to provide order within society, there are instances when social rules are not followed. An act or belief which violates a group's norms is referred to as **deviance**. Outside of sociology, the term deviance is typically associated with crime or negative behaviors. However, within sociology, deviance is defined more broadly to include uncommon (eg, eating a vegan diet) or off-putting (eg, nose-picking in public) acts, as well as egregious (eg, marital infidelity) or criminal (eg, murder) activities.

Norms, and therefore what is considered deviant, are social constructs (Concept 32.2.04). Definitions of deviant behavior are specific to the social context and vary across time periods. Behaviors that are considered deviant or even criminal in one setting (eg, killing a family member is constructed as murder) may be perfectly acceptable in another (eg, killing enemies during a war is constructed as heroic).

42.2.02 Perspectives on Deviance

In sociology, there are a variety of perspectives used to explain why deviance occurs. Each theory on deviance describes why deviant acts and beliefs are a part of social life despite the importance of normative behavior to maintain order and stability in society.

Differential association theory suggests deviance is learned through interaction with others engaging in deviance. In this theory, socialization into groups is important to understand whether an individual engages in normative or deviant behaviors. At times, a group may engage in behaviors that are considered deviant within mainstream society but within the group the behavior is common. For example, if gang members carry illegal firearms, new recruits quickly learn that this is "normal" and that they are expected to engage in this deviant behavior to be a part of the group.

Another perspective on deviance, **labeling theory**, proposes that deviance lies not in the behavior but in the social response of applying the "deviant" label to individuals. When someone is labeled as deviant, the act of being labeled produces further deviance. The initial act is called **primary deviance** and is usually mild but leads to the "deviant" label and resulting social stigma (ie, disapproval by others). Internalization of the deviant label leads to further and often more serious violations called **secondary deviance**.

For example, if an individual is convicted of a crime and labeled a criminal, they may experience stigmatization when seeking employment and housing, and police may be more likely to monitor the individual's activities. These obstacles facilitate the individual's internalization of the label "criminal," making it more likely that person will commit another crime and be caught again. Therefore, the response to behaviors (ie, labeling and social stigma), not the initial behaviors themselves, results in further acts of deviance.

Finally, **strain theory** argues that deviant behavior results from tension (ie, strain) caused by a disconnect between socially acceptable **goals**, such as a career and middle-class lifestyle, and the **means** to obtain those goals, such as access to higher education. For example, a parent is unable to feed her child (goal) because she does not have enough money (lack of means) and therefore experiences strain. This strain may then cause her to seek deviant means (eg, stealing food) to achieve the goal.

The three key perspectives on deviant behavior are summarized in Table 42.2.

Table 42.2 Perspectives on deviant behavior.

Differential association	Individuals learn specific deviant behaviors and values/norms through interaction with others with those same behaviors and values/norms.
Labeling	Primary deviance (a small social norm violation) leads to a deviant label and social stigma, causing secondary deviance (more serious violations).
Strain	Deviant behavior results from the disconnect between goals and the means for achieving those goals.

END-OF-UNIT MCAT PRACTICE

Congratulations on completing **Unit 8: Identity and Social Interaction**.

Now you are ready to dive into MCAT-level practice tests. At UWorld, we believe students will be fully prepared to ace the MCAT when they practice with high-quality questions in a realistic testing environment.

The UWorld Qbank will test you on questions that are fully representative of the AAMC MCAT syllabus. In addition, our MCAT-like questions are accompanied by in-depth explanations with exceptional visual aids that will help you better retain difficult MCAT concepts.

TO START YOUR MCAT PRACTICE, PROCEED AS FOLLOWS:

1) Sign up to purchase the UWorld MCAT Qbank
   IMPORTANT: You already have access if you purchased a bundled subscription.
2) Log in to your UWorld MCAT account
3) Access the MCAT Qbank section
4) Select this unit in the Qbank
5) Create a custom practice test

Unit 9 Demographics and Social Structure

Chapter 43 Social Institutions

43.1 Education

43.1.01	Hidden Curriculum	
43.1.02	Teacher Expectancy	
43.1.03	Educational Segregation and Stratification	

43.2 Family

43.2.01	Kinship	
43.2.02	Diversity in Family Forms	
43.2.03	Abuse in the Family	

43.3 Religion

43.3.01	Religiosity	
43.3.02	Types of Religious Organizations	
43.3.03	Religion and Social Change	

43.4 Government and Economy

43.4.01	Power and Authority	
43.4.02	Division of Labor	

43.5 Medicine

43.5.01	Medicalization	
43.5.02	The Sick Role	
43.5.03	The Illness Experience	
43.5.04	Social Epidemiology	

Chapter 44 Demographic Structure of Society

44.1 Age

44.1.01	Aging and the Life Course	
44.1.02	Age Cohorts	

44.2 Gender

44.2.01	Sex versus Gender	
44.2.02	Gender Segregation	

44.3 Sexual Orientation

44.3.01	Sexual Orientation	

44.4 Race and Ethnicity

44.4.01	Social Construction of Race and Ethnicity	
44.4.02	Racial Formation and Racialization	

44.5 Immigration Status

44.5.01	Patterns of Immigration	

Chapter 45 Social Class and Inequality

45.1 Social Stratification

45.1.01	Social Class and Socioeconomic Status	

      45.1.02   Class Consciousness and False Consciousness
      45.1.03   Cultural Capital and Social Capital
      45.1.04   Social Reproduction
      45.1.05   Power, Privilege, and Prestige
      45.1.06   Intersectionality
      45.1.07   Socioeconomic Gradient in Health

45.2 Social Mobility

      45.2.01   Intergenerational and Intragenerational Mobility
      45.2.02   Vertical and Horizontal Mobility
      45.2.03   Meritocracy

45.3 Spatial Inequality

      45.3.01   Residential Segregation
      45.3.02   Neighborhood Safety and Violence
      45.3.03   Environmental Justice

45.4 Poverty

      45.4.01   Absolute and Relative Poverty
      45.4.02   Social Exclusion

45.5 Health and Healthcare Disparities

      45.5.01   Health Disparities
      45.5.02   Healthcare Disparities

Chapter 46 Social Change

46.1 Urbanization

      46.1.01   Urbanization

46.2 Globalization

      46.2.01   Globalization

46.3 Social Movements

      46.3.01   Social Movements

46.4 Demographic Change

      46.4.01   Theories of Demographic Change
      46.4.02   Population Growth and Decline
      46.4.03   Fertility and Mortality
      46.4.04   Push and Pull Factors in Migration

Lesson 43.1
Education

43.1.01 Hidden Curriculum

Social institutions (eg, family, religion) are enduring, organized systems that outline behavioral norms and fulfill a specific purpose in society. The persistence of social institutions over time helps to create stability and predictability in social life. The **education system** is a social institution that meets society's need to transfer knowledge and skills to its members.

Classroom interactions and experiences are a part of socialization (see Lesson 34.4). The education system is designed to teach a formal curriculum (ie, explicit, official content) such as reading and math. Additionally, the education system has a latent function (ie, unintended purpose which is described in Concept 32.2.01) to transfer dominant cultural values. The **hidden curriculum** includes the implied, informal mechanisms by which certain beliefs and behaviors are promoted within academic settings. For example, reciting the national anthem at the start of each school day is a part of the hidden curriculum instilling the value of patriotism.

43.1.02 Teacher Expectancy

Student performance is impacted by **teacher expectancy** when a teacher's beliefs about a student (eg, a particular student is "smart" or "lazy") result in the student meeting those expectations (eg, excelling or performing poorly, respectively). Like self-fulfilling prophecy (see Concept 40.2.02), teacher expectancy is thought to occur because teachers' beliefs unconsciously influence their actions, causing them to treat students differently (eg, giving less feedback to a student believed to be "lazy"). Students may then overperform when the teacher has high expectations or underperform when the teacher has low expectations. This process is depicted in Figure 43.1.

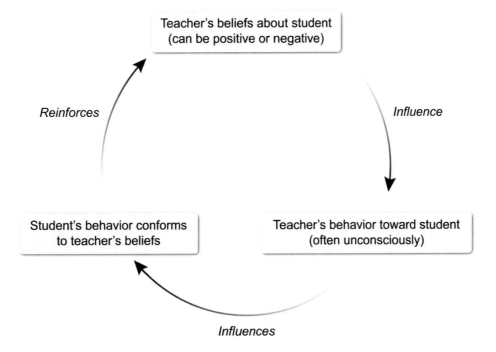

Figure 43.1 Impact of teacher expectancy on student outcomes.

43.1.03 Educational Segregation and Stratification

Educational stratification refers to the mechanisms that produce inequality in educational access (eg, the schools that are available to students) and outcomes (eg, graduation rates, college matriculation) in society. Students are stratified in the education system because of social inequalities related to race, class, and gender. For instance (see Figure 43.2), students with high social standing have many options (eg, well-funded local public schools, private schools), whereas students with low social standing have fewer options (eg, limited access to quality schools and teachers).

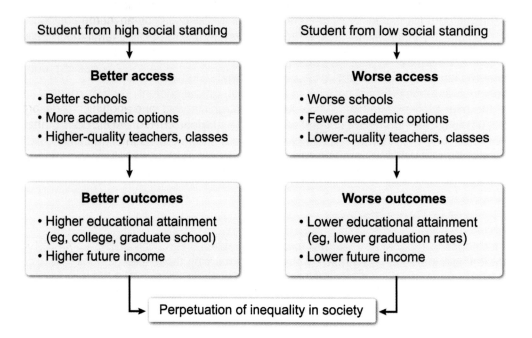

Figure 43.2 Educational stratification and the impact on social inequality.

Lesson 43.2
Family

43.2.01 Kinship

The family is a social institution designed to meet basic human needs, including food, shelter, and support. **Family** is broadly defined as at least two individuals connected by an intimate relationship (eg, marriage/partnership, birth, adoption) who often live together. The definition of family varies based on culture (eg, who counts as a family member) and can include different types of bonds between individuals, including intimate/emotional bonds (eg, romantic partnership), legal bonds (eg, adoption, marriage), and/or genetic bonds (ie, biological family).

In sociology, **kinship** describes how individuals in society are related to one another. There are three major types of kinship (see Table 43.1): **consanguineal** (genetically related individuals), **affinal** (individuals related through marriage), and **fictive** (individuals related through something other than genetics or marriage). Also known as chosen or voluntary kin, fictive kin may share various types of bonds, including those through law (eg, adopted children), religion (eg, godparents), and close friendships (eg, fraternity brothers).

Table 43.1 Types of kinship bonds.

Kinship bond	Definition
Consanguineal	Based on a genetic relationship (eg, biological parents)
Affinal	Based on marriage (eg, spouses)
Fictive	Based on social ties that are not consanguineal or affinal (eg, godparents)

43.2.02 Diversity in Family Forms

Family forms describe which individuals are included in the definition and structure of a family group. In Western societies during the post-World War II period, the "typical" family form consisted of a breadwinner (ie, economic earner) male parent and a homemaker (ie, caretaker of children and home) female parent with biological children. In society today, there is greater diversity in family forms with many alternative combinations of family members. Some examples of diverse family forms include heterosexual or same-sex couples with or without children, single parents, couples with adopted children, couples in second marriages with step-children, and grandparents with grandchildren.

43.2.03 Abuse in the Family

While the function of the family is to provide support and stability for its members, unfortunately, abuse can occur. Abuse within families can take multiple forms, including physical or sexual violence, emotional harm, neglect, and/or isolation.

Cases of abuse in families are categorized based on family relationships and the age of the victims. **Child abuse** refers to the mistreatment and neglect of children; abuse experienced in early life can have lifelong consequences for the victims. **Spousal abuse** (also known as domestic violence or intimate partner violence) includes harassing and/or harmful behaviors toward one's partner. **Elder abuse** is a more recently defined category of abuse within families and can include financial abuse (ie, mishandling or stealing an elder's money).

Lesson 43.3
Religion

43.3.01 Religiosity

Religion is a social institution that addresses the spiritual needs of a society and can provide a framework for learning norms and values. In sociology, religious groups such as Islam, Hinduism, and Christianity are defined as cultural systems (see Lesson 34.1). Each religion's specific beliefs and rituals support the group's understanding of morality (ie, right/wrong) and the meaning of life/death.

Furthermore, an individual's involvement in religious groups impacts their identity. **Religiosity** (also known as religiousness) refers to the extent to which a given religious doctrine (ie, set of beliefs and principles) is incorporated into all aspects of an *individual's life*. In other words, religiosity is the degree to which a person internalizes a religion. For example, those who consider themselves Jewish and have a high degree of religiosity demonstrate beliefs and behaviors that align with Judaism, and "being Jewish" is an important part of their identity.

43.3.02 Types of Religious Organizations

The many different religious organizations that exist worldwide can be divided into three main categories: churches, sects, and cults (see Table 43.2). **Churches** are large, formal organizations with traditional practices and often include bureaucratic leadership (eg, the hierarchy of priest, bishop, and pope in Catholicism) (see Concept 36.6.01 for formal organizations and Concept 36.6.02 for bureaucracies). In sociology, a "church" is not limited to Christian faith communities; any religious organization that is mainstream and well-established in society is categorized as a church, including Jewish, Buddhist, and Islamic religions.

Some religious organizations are not mainstream or integrated into society. **Sects** are religious subgroups formed after believers split from an established church in pursuit of a more traditional form of faith. Some sects utilize more equal and open rather than hierarchical leadership. Members of sects aim to revive the "true beliefs" of a religion that have been diminished or corrupted. For example, Hasidic Judaism is a sect that promotes a return to traditional Jewish beliefs and practices including religious rituals and style of dress.

New religious organizations are categorized as **cults**. These new groups present claims about spirituality that are not present in mainstream religions, and many cults are loosely organized around a charismatic (ie, inspirational and empowering) leader. The term "cult" can be negative; however, sociologists use this term to describe *any* new religious group, without evaluating or judging the group.

Table 43.2 Types of religious organizations.

	Organizational structure	Relation to society
Church	Large, formal organization with bureaucratic leadership	Mainstream and established religious group
Sect	Smaller, informal organization often with equal and open leadership	Subgroup of an established religious group
Cult	Loose organization around a charismatic leader	New group offering a new religious belief system

The types of religious organizations operate on a spectrum, meaning groups that are currently classified as sects or cults may eventually be categorized as churches. For example, today, Protestant groups such as Lutherans and Methodists are considered churches but were originally small sects protesting the corrupt practices within the Catholic church during the 1500s in Europe.

43.3.03 Religion and Social Change

As elements of society change, there is an impact on the practice and purpose of religious organizations (see Table 43.3). **Modernization** refers to the social progress and transition of a society originally brought about by industrialization (ie, development of systems of production focused on efficiency). Over time, this process results in a society becoming less traditional with new cultural values. Within religion, modernization marks a shift away from traditional practices to better reflect the modern world. For example, the incorporation of female clergy into historically male-dominated religious groups demonstrates the impact of modernization on religion.

The focus on efficiency brought about by modernization relies on rational, scientific ideas. The **secularization** of society refers to a shift from traditional ways of life to a society organized by science and empiricism (ie, observable phenomena). Within religion, this process diminishes the social and political power of religious organizations. For example, the shift from traditional religious healing practices to the use of scientific medical interventions to treat illness illustrates secularization. Secularization occurs at the micro-level with less individual involvement in religious activities and at the macro-level with a decreased influence of religious institutions on other aspects of social life (eg, education).

In reaction to modernization and secularization, **fundamentalism** is the renewed adherence to strict, traditional beliefs and practices. Within religion, fundamentalist groups often believe in a literal interpretation of religious doctrine (eg, sharia, or traditional Islamic law) and texts (eg, the Torah, the Bible). The strict interpretation of religious beliefs and practices sometimes leads to intolerance of others (eg, violence towards non-believers).

Table 43.3 Impact of social change on religion.

Process of social change	Impact on religion
Modernization	Changes traditional religious practices to fit into the modern world
Secularization	Reduces the power of religion as religious involvement declines
Fundamentalism	Renews commitment to traditional religion as a reaction to secularization

Lesson 43.4

Government and Economy

43.4.01 Power and Authority

The government and economy are institutions that function at a variety of levels within society (eg, local, state, national). **Governments** are legal institutions that provide order and stability through social services (eg, postal, transportation, public health) and the implementation of laws. The **economy** regulates the production, distribution, and consumption of commodities and services throughout society.

While there are different systems of government (eg, monarchy, democracy) and economy (eg, capitalist, socialist) across the world, one common feature is the reliance on power for the systems to function effectively. **Power** is the ability of individuals or groups to act based on their own interests to achieve goals with little resistance. Institutions like the government and economy must have the power to influence the societal order and maintain stable operations. Additionally, many types of social relationships involve differences in power, such as manager/employee, physician/patient, and parent/child.

In some instances, power is exerted by force and coercion (eg, in a violent government takeover). In contrast, the legitimate and justified application of power is referred to as **authority**. Individuals subject to authority believe in the validity of this form of power. For example, citizens recognize the court as having legitimate power to limit individual freedoms such as the loss of a driver's license as punishment for driving while intoxicated. Sociologist Max Weber described three types of authority:

- **Traditional authority** comes from longstanding patterns in society. For example, a royal family bloodline is seen as having legitimate power due to the traditions of a monarchy.
- **Charismatic authority** stems from an individual's personal appeal and/or extraordinary claims. For example, Gandhi was seen as having legitimate power due to his ability to inspire people.
- **Rational-legal authority** arises from a person's professional position. For example, a physician is seen as having legitimate power due to their extensive training.

43.4.02 Division of Labor

Government and economic institutions rely on the **division of labor** which refers to the specialization of tasks in society. Current economic systems often approach large tasks (eg, manufacturing a product) by separating work based on skill and specialized training. Through the division of labor, societies become more interdependent, and citizens rely on one another to accomplish the tasks needed for survival.

For example, in a low division of labor society, individuals engage in subsistence farming and produce all the food their family will eat. In a high division of labor society, food production is divided among multiple individuals. A farmer grows and harvests the food, another person transports the food to a store, and then store workers sell the food to consumers. Each step has a specialized task, and multiple individuals need to complete their tasks for food to be available for everyone to eat.

Division of labor has advantages and disadvantages. When labor is specialized and divided among multiple individuals, there is an increase in efficiency and productivity along with a reduction of costs (eg, assembly line production). However, specialization can also result in the exploitation of labor when workers' rights are diminished in pursuit of productivity and profit (eg, workers' breaks are eliminated to increase productivity).

Similarly, specialization can contribute to inequality in society when certain tasks and occupations are seen as more valuable; however, this valuation does not always reflect how necessary a task is to society's functioning. For example, a garbage collector is a necessary occupation in society, but this specialization has low value. However, a professional athlete is a less necessary occupation in society, but this specialization has high value.

Lesson 43.5
Medicine

43.5.01 Medicalization

Medicine is the social institution responsible for the promotion and maintenance of health within society. Definitions of health and approaches to healing differ between cultural groups. For example, Traditional Chinese Medicine views health as a balance of Qi (ie, vital life force), and healing practices such as acupuncture and herbs aim to restore balance in the body. However, in Western societies biomedicine (ie, science-based medicine) views health as the absence of disease, and physicians utilize medical technologies (eg, MRI, CT scan) and pharmaceuticals to diagnose and treat symptoms, respectively.

Contributing to the social significance of biomedicine is **medicalization**, which is the process of defining human behaviors or characteristics as medical conditions. Once a condition is defined as a medical issue, medical professionals are consulted in diagnosing, preventing, or treating the condition. For example, menopause, a biological experience associated with natural aging, is now viewed by some as a deficiency in hormones that can be treated with pharmaceuticals.

Positive outcomes of medicalization include treatment, awareness, and funding for certain conditions. For example, Vietnam War veterans who were experiencing disturbing memories and flashbacks of their trauma advocated for the medicalization of their condition through the development of the posttraumatic stress disorder (PTSD) diagnosis, which helped the veterans gain access to necessary treatment. On the other hand, negative consequences of medicalization include side effects of medication (eg, hormone treatment for menopause increases risk of stroke and breast cancer) and stigmatization (ie, negative labeling of individuals as described in Concept 40.4.03).

43.5.02 The Sick Role

In sociology, **sick role theory** is a functionalist perspective (Concept 32.2.01) that describes how the disruptions to typical social activity (eg, work, school) are minimized through the "sick role." When ill, an individual follows the social expectations of the "sick role," including designated rights and obligations:

- **Rights**: The sick person has the right to be exempt from performing other social roles (eg, employee) while sick and is excused from fulfilling normal responsibilities (eg, going to work). The sick person also has the right to not be blamed for the illness.
- **Obligations**: The sick person has the responsibility to make every reasonable effort to get well as soon as possible. The sick person also has the responsibility to seek medical help and to cooperate with medical professionals (eg, follow their treatment plan).

The rights and obligations of the sick role help maintain order within society and guide sick individuals to restore their health and return to their typical responsibilities. While being sick may be considered a non-normative behavior (Concept 42.2.01), the application of the sick role legitimizes illness as a socially acceptable form of deviance.

43.5.03 The Illness Experience

The **illness experience** is a symbolic interactionist approach (Concept 32.2.03) that examines how illness impacts identity and daily life. In sociology, "illness" is a subjective interpretation of sickness and health, whereas "disease" is objectively defined by medical professionals. For example, an individual who has celiac disease (ie, gluten intolerance) may subjectively interpret themselves as healthy and not experiencing *illness* if they manage their symptoms through diet change.

The illness experience also addresses the fact that individuals with chronic illness must make sense of and manage their illness in everyday life. For example, these individuals typically gather information about their illness and seek necessary treatments. Chronic illness can impact interaction, such as explaining the illness to coworkers, friends, and family. Additionally, daily activities that do not directly involve managing or explaining one's illness are typically nevertheless impacted, such as maintaining a household, working, and caring for loved ones.

43.5.04 Social Epidemiology

To understand the impact of health and illness on society, public health professionals and other scientists research patterns of disease and illness experiences. Epidemiology is the study of disease incidence (ie, new cases of a disease) and prevalence (ie, proportion of the population with a disease); **social epidemiology** is a subfield of epidemiology that focuses on the social factors that impact the health of a population. Social factors impact health outcomes on many levels, including structural factors (eg, a society at war), cultural factors (eg, school policies on vaccination), and individual factors (eg, childhood exposure to toxins).

Lesson 44.1
Age

44.1.01 Aging and the Life Course

One way sociologists understand the organization of society is through **demographics** (ie, statistical analysis of patterns and trends occurring within certain categories, such as age, gender, race/ethnicity, and sexual orientation). For example, the U.S. Census is completed every decade to collect and compare demographic data about the citizens of the United States.

Aging involves changes that are biological (ie, physical changes to the body, such as decreased metabolism and organ function), psychological (eg, changes in cognitive abilities, such as slower processing speed), and social (eg, changes in social roles, such as becoming a retiree). Additionally, the social importance and purpose of life stages can differ between societies. For example, many Asian societies value the wisdom of the elderly over youth, whereas in Western societies youth is highly valued.

The **life course approach** is a holistic and multidisciplinary examination of aging in terms of psychological, biological, and sociocultural factors across a lifetime. This approach emphasizes the reciprocal link between social context and individuals as they age (ie, individuals influence and are influenced by society). The life course approach considers how personal life events (eg, illness in infancy), individual choices/behaviors (eg, having unprotected sex), and sociocultural and historical context (eg, being born during wartime) impact health and illness.

44.1.02 Age Cohorts

Individuals who were born around the same time have shared sociocultural and historical experiences (eg, growing up with the internet at home); therefore, categorizing individuals based on age range can be useful to understand patterns within society. **Age cohorts** refer to groups of individuals born within a specified time frame (eg, individuals born between 2000 and 2010). Generation categories are one type of age cohort; Table 44.1 describes four of the largest current generations in the United States.

Table 44.1 Some current generations by birth year range.

Generation	Birth year range	Example of shared experiences
Baby Boomers	1946–1964	Childhood during the post-World War II era
Generation X	1965–1980	Entered adulthood after the civil rights movement
Millennials (Generation Y)	1981–1996	Old enough to understand the global impact of the 9/11 terrorist attack
Generation Z	1997–2012	Grew up with digital technology and social media

Studying patterns among age cohorts can help sociologists understand changes in demographic trends. For example, the majority of Baby Boomers, a large age cohort in the U.S. population, have reached retirement age and many are no longer active in the labor force. This demographic trend has an impact on other aspects of society, including the government (eg, increased withdrawals from Social Security) and healthcare (eg, increased use of Medicare insurance).

Lesson 44.2
Gender

44.2.01 Sex versus Gender

Although some people use the terms "sex" and "gender" interchangeably, behavioral scientists distinguish between the two concepts. **Sex** refers to the biological distinction between male and female bodies typically assigned at birth, whereas **gender** refers to cultural beliefs about which behaviors (eg, wearing suits or dresses) and roles (eg, financial provider or caregiver) are considered masculine or feminine.

In sociology, gender is defined as a social construct (ie, meaning is created through social interaction as described in Concept 32.2.04). Cultural groups have varying definitions of masculinity and femininity which create differences in gender behaviors and norms. For example, religious values like modesty can influence expected behaviors related to gender such as women wearing long dresses to conceal most of the body.

Gender impacts how one behaves, views themselves, and is treated by others. **Gender identity** refers to how someone sees themself in relation to the cultural definitions of gender (eg, man/woman), and **gender roles** refer to the expected behaviors and attributes associated with masculinity and femininity (see Table 44.2). Historically, gender has been defined as binary categories (ie, behavior or identity was understood as either masculine or feminine). However, recent cultural changes have led to the reconceptualization of gender as a continuum rather than a binary which results in greater variation in gender identities (eg, nonbinary, gender-fluid) and behaviors.

Table 44.2 Key concepts related to sex and gender.

Concept	Definition
Sex	Biological differences between males and females (eg, physical differences)
Gender	Cultural norms and beliefs regarding masculinity and femininity
Gender identity	One's sense of self in relation to gender (eg, man, woman, gender-fluid)
Gender role	Behaviors and attributes society considers appropriate based on gender

44.2.02 Gender Segregation

One mechanism that maintains gender differences is **gender segregation,** which refers to the separation of individuals based on their assumed gender. This separation can occur by formal means (eg, policies about gendered bathrooms) or informal means (eg, social norms about gendered behaviors). Gender stereotypes about masculine and feminine behaviors can reinforce gender segregation. For example, the stereotype that men are more rational than women may influence gender segregation in leadership (such as the practice of appointing men as leaders rather than appointing women).

Gender segregation can create differences in status and limit access to resources, contributing to **gender inequality** (ie, gender hierarchy which places greater value on masculinity over femininity). For example, sports teams are often segregated based on gender. If there are gender differences in funding, facilities, and opportunities (eg, men's teams receive greater support), gender segregation in sports could reinforce gender inequality.

Lesson 44.3
Sexual Orientation

44.3.01 Sexual Orientation

In sociology, sexuality refers to one's thoughts, feelings, and behaviors as a sexual being. One aspect of sexuality is **sexual orientation**, which describes a person's identity based on their attraction to others. Historically, sexual orientation was divided into two distinct categories, with *homosexuality* describing attraction to members of one's own sex and *heterosexuality* describing attraction to members of the opposite sex. Today, sexual orientation is described using a greater diversity of categories including homosexual, heterosexual, bisexual (attracted to both sexes), and asexual (little or no sexual attraction) identities.

Sexual orientation can affect how an individual is treated by others (eg, discriminated against, welcomed in a community) as cultures vary in their acceptance of differing sexual orientations. For example, in some societies, there is a value of heteronormativity (ie, heterosexuality is seen as the typical and only acceptable orientation), which results in the limitation of rights and the mistreatment of individuals with other sexual orientations.

Lesson 44.4
Race and Ethnicity

44.4.01 Social Construction of Race and Ethnicity

Race is a demographic category that groups individuals based on shared physical traits (eg, hair texture, skin color) and a presumed shared ancestry (ie, bloodline). While race was previously assumed to reflect biological differences between groups of people, there are no clear genetic markers that characteristically define any racial category. Currently, the U.S. Census Bureau utilizes five racial categories: Black or African American, American Indian or Alaska Native, Asian, Native Hawaiian or other Pacific Islander, and White.

Despite there being no conclusive biological basis for racial categories, race remains a meaningful social category, impacting one's position in society (Lesson 36.2) as well as the development of one's identity (Lesson 35.2). Racial categories are considered social constructs (Concept 32.2.04) because definitions of race have changed throughout history and vary by society. For example, an individual with one Black parent and one White parent may be considered Black in the United States but considered mixed-race in another country.

Another demographic category connected to ancestry is **ethnicity**, which refers to groups of people who share cultural beliefs, language, and history. Ethnicities are also social constructs, and the definition and significance of ethnic groups can vary. For example, in the United States during the nineteenth century, ethnicities such as Irish American, Polish American, and Italian American were considered important categories that signaled cultural differences between the ethnic groups. However, in the United States today, these ethnic categories have less significance and many individuals in these ethnic groups have assimilated into mainstream U.S. culture (Concept 34.3.03).

44.4.02 Racial Formation and Racialization

According to the **racial formation theory** developed by sociologists Michael Omi and Howard Winant, historical, political, and social contexts need to be taken into account to understand current race relations; for example, the history of colonialism and slavery in the United States must be considered when examining the current experiences of discrimination and inequality for African Americans. Additionally, racial formation theory posits that race is a part of all social relations (eg, personal connections, social organizations), and perceptions of an individual's racial identity impact how they are treated by others.

Within racial formation theory, the concept of **racialization** describes the process whereby one group designates another group with a racial identity. The designating group has more social power and exerts social control over the designated group, creating a racial hierarchy. For example, in the United States, racialization occurred during colonialization: European settlers created the new racial category "Indian" to group indigenous (ie, native) people together, which led to social control of the group through policies such as the Indian Removal Act (ie, legal displacement of tribes to allow for westward expansion).

Lesson 44.5
Immigration Status

44.5.01 Patterns of Immigration

Immigration is the relocation of individuals into an area, whereas **emigration** is relocation out of an area. People tend to emigrate from less developed or poorer countries, immigrating to more developed or wealthier ones. **Immigration status** is a demographic category that refers to an individual's citizenship status, including lawful permanent resident, refugee, and unauthorized immigrant.

Federal policies can shape patterns of immigration and alter the demographics of a nation. For example, a policy in the 1960s ended the national quota system (which had limited the number of immigrants allowed to enter the United States per country of origin), contributing to an increase in the total immigrant population in the United States (Figure 44.2). Based on data collected by the U.S. Census Bureau, most immigrants currently live in the West and South regions of the country, and the largest proportion of foreign-born residents emigrate from Mexico, China, and India.

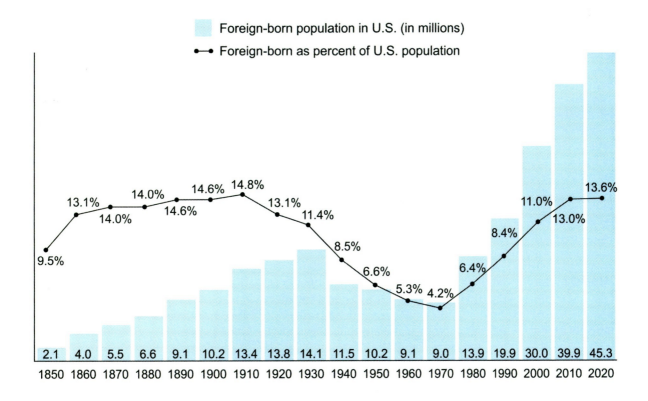

Figure 44.2 United States immigrant population from 1850–2020.

Lesson 45.1

Social Stratification

45.1.01 Social Class and Socioeconomic Status

In sociology, **social stratification** refers to a system of inequality whereby individuals are hierarchically ranked into groups based on social identity categories (eg, age, race, gender) that have differential access to resources, opportunities, and life outcomes. Caste and class systems are two types of social stratification systems present in societies today.

In a **caste system** (eg, India), an individual's social position (ie, location within the social hierarchy) is involuntarily assigned at birth based on personal characteristics such as race, ethnicity, or parental caste. Individuals tend to remain in the same caste for life and are discouraged (formally and informally) from intimate contact with individuals in other castes. Therefore, movement between castes is very difficult.

In a **class system** (eg, the United States), an individual's social position is partially achieved through effort and/or ability and partially based on the social class (see definition below) into which one is born. Social class positions are less clearly defined as compared to castes, and boundaries between classes are not definite. Therefore, movement between classes is possible.

A **social class** (also known as socioeconomic status or SES) is a group of people who share similar lifestyles based largely on economic resources (eg, income, wealth, property). There are three generally recognized classes in the United States: the upper, middle, and lower classes (as summarized in Table 45.1).

Several additional factors contribute to a person's social class, including the SES of one's parents, educational attainment, and occupation. Social class impacts life outcomes and opportunities. For example, individuals in the upper class have the most resources (eg, access to the best schools, healthcare, and jobs) and power (ie, ability to influence or control others as defined in Concept 45.1.05).

Table 45.1 Social classes in the United States.

Social class	Description	Approximate proportion
Upper class	Wealthiest, most influential members of society (eg, heirs to family fortunes, top executives)	21%
Middle class	Educated professionals with average to high wages (eg, teachers, lower-level management, nurses, doctors, lawyers)	50%
Lower class	Clerical, low-skill service, and manual laborers with low (ie, minimum wage) to adequate wages (ie, living wage) (eg, factory workers, clerical workers, day laborers, service industry)	29%

45.1.02 Class Consciousness and False Consciousness

In Karl Marx's theory of society, the capitalist economic system relies on two unequal classes (ie, bourgeoise/upper class and proletariat/lower class; see Concept 32.2.02). **Class consciousness** involves the recognition of the class structure and identification with one's own social class position. In other words, class consciousness allows individuals to understand that people from their class have shared needs and interests while the goals of other classes differ. According to Marx, class consciousness is needed for the lower class to engage in collective political action to challenge the unequal capitalist economic system.

Marx also asserted that individuals in the upper class, to further their own interests, attempt to sway the thinking of those in the lower class. **False consciousness** results when individuals from lower classes adopt misleading views of the class structure and accept the status quo (eg, injustice, exploitation). For example, individuals from the lower classes who do not have class consciousness may blame themselves for living in poverty rather than understanding the impact of the unequal class system on their life outcomes. Consequently, false consciousness results in class inequalities remaining unchallenged.

45.1.03 Cultural Capital and Social Capital

Sociologists define **capital** as something possessed by an individual that confers advantage in society. Capital can be accumulated (ie, compiled) and applied to achieve desired outcomes, such as increased socioeconomic status.

There are three major types of capital (see Figure 45.1). **Economic capital** describes an individual's tangible financial assets, such as property and income. **Social capital** refers to benefits associated with an individual's social networks (see Concept 36.5.02). In other words, the people an individual knows can provide advantages (eg, knowing the manager helps an individual get the job). **Cultural capital** describes how knowledge of cultural practices and skills can provide access to power and opportunity in society. For example, when a business meeting is held at a formal restaurant, an employee with knowledge of fine dining etiquette has cultural capital that provides an advantage in this setting (ie, others may favorably evaluate the employee who uses appropriate fine dining skills).

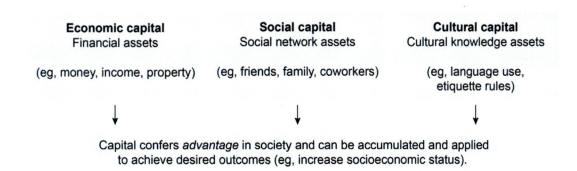

Figure 45.1 Three major types of capital.

45.1.04 Social Reproduction

Social stratification systems remain consistent over time. **Social reproduction** refers to the transmission of society's values, norms, and practices, including social inequality, from one generation to the next. Social reproduction occurs because social structures (eg, tax laws, education systems) maintain and perpetuate social inequality over time; therefore, successive generations tend to occupy the same social class (eg, the child of wealthy parents tends to be wealthy as an adult).

45.1.05 Power, Privilege, and Prestige

In sociology, the interconnected concepts of power, privilege, and prestige help explain how systems of stratification function:

- **Power** is the ability to control others and act based on one's interests. Certain careers (eg, corporate executive) and accomplishments (eg, earning a professional degree) increase one's power.
- **Privilege** describes the advantages and opportunities associated with social positions. Certain social identity categories (Concept 35.1.02) receive unearned benefits based on the organization of society (eg, growing up in an upper-class neighborhood confers the educational privilege of well-funded schools through property tax).
- **Prestige** refers to one's positive reputation or revered attributes, compared to others. Certain careers (eg, politicians), personal characteristics (eg, ambitiousness), and achievements (eg, being admitted to a top-tier medical school) confer prestige.

Power, privilege, and prestige operate to maintain an unequal distribution of resources, opportunities, and life outcomes. For example, a male CEO has *power*, and he can implement a policy that reduces family leave, which supports his interests and controls the behavior of others. The policy change also confers *privilege* to male workers because men are typically less responsible for family care and therefore less likely affected by the policy. Finally, the CEO may increase his *prestige* in the eyes of the company owner by developing a policy that increases profits.

45.1.06 Intersectionality

Intersectionality describes how individuals hold multiple, interconnected social identities (eg, gender, race, age) that simultaneously impact their perspectives and treatment in society. Each social identity category is associated with different privileges and/or disadvantages.

An individual does not interact with others solely based on race, gender, or any other social identity alone and therefore can experience more than one type of discrimination (eg, racism, ageism). For example, a 50-year-old, middle-class, Black woman (as shown in Figure 45.2) will have different experiences than a 25-year-old, upper-class, White woman. While both women may experience gender discrimination, the intersections of race, class, and age also impact how they are treated by others, resulting in differences in life outcomes.

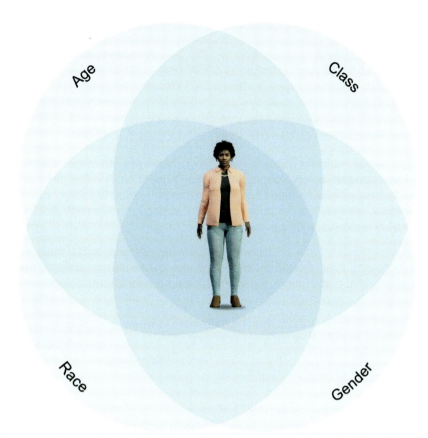

For example, an individual with multiple, interconnected social identities that include being a 50-year-old, middle-class, Black woman reflects intersectionality.

Figure 45.2 Example of intersectionality.

45.1.07 Socioeconomic Gradient in Health

The **socioeconomic gradient in health** is the positive correlation (see Concept 3.2.01) between socioeconomic status (SES) and health outcomes. Individuals with lower SES tend to have worse health outcomes than those with higher SES, on average. Disparities (ie, inequalities) related to SES are thought to contribute to this gradient, including lower income, poorer education, unsafe living environments, lack of access to healthcare, lack of exercise/leisure time, and limited access to nutritious food.

Lesson 45.2
Social Mobility

45.2.01 Intergenerational and Intragenerational Mobility

In sociology, the term **social mobility** refers to the change or movement of individuals, groups, or families within a social stratification system (eg, from middle class to upper class). Social mobility can be horizontal, upward, or downward (Concept 45.2.02) and can be related to many other factors, such as educational achievement, job loss, career advancement, marriage, and/or institutional discrimination (Concept 40.3.02 covers institutional discrimination).

If a group or person experiences social mobility within their own lifetime, social mobility is said to be **intragenerational**, occurring over a relatively short time within a single generation. For example, if an individual grew up in a middle-class family and then marries someone from an upper-class family, intragenerational mobility has occurred.

Social mobility that takes place over more than one life span is **intergenerational**, occurring over a relatively long time among members of different generations. For example, if parents do not have college degrees and are in the lower class, but their child earns a college degree and has a successful middle-class career, intergenerational mobility has occurred.

45.2.02 Vertical and Horizontal Mobility

Movement within systems of stratification (Lesson 45.1) can occur in a variety of directions. **Vertical social mobility** refers to movement up or down the social class hierarchy. *Upward mobility* occurs when an individual advances or moves up from one class to a higher class through changes in occupation, educational level, or marriage. *Downward mobility* occurs when an individual descends to a lower class and can be caused by large-scale changes in the economy (eg, national recession) or by individual events (eg, being fired from a job).

Alternatively, **horizontal social mobility** refers to movement within the same level of the social class hierarchy; therefore, one's power and opportunities remain the same in this type of change. For example, an individual could change specialized occupations within the automotive repair industry, such as switching from working as a mechanic to an electronics installer, and this movement would not change the individual's social class position.

Intragenerational and intergenerational mobility (Concept 45.2.01) can occur horizontally and vertically. Table 45.2 illustrates the various combinations of time frame (ie, generations) and direction of movement (ie, none, increase, or decrease).

Table 45.2 Types of social mobility.

	Direction of movement		
Time frame of movement	**No change in socioeconomic status** ←→	**Increase in socioeconomic status** ↑	**Decrease in socioeconomic status** ↓
1 Generation	Intragenerational horizontal mobility (eg, doctor moves her practice from Atlanta to Chicago)	Intragenerational upward mobility (eg, after 10 years, nurse goes back to school to earn MD)	Intragenerational downward mobility (eg, hospital closes and doctor cannot find another job)
2+ Generations	Intergenerational horizontal mobility (eg, son of doctors becomes a doctor)	Intergenerational upward mobility (eg, son of high school dropouts becomes a doctor)	Intergenerational downward mobility (eg, son of doctors becomes a high school dropout)

45.2.03 Meritocracy

A **meritocracy** is a system in which an individual's socioeconomic status (SES) is based on personal abilities and hard work. In other words, factors such as prestige (ie, one's positive reputation, see Concept 45.1.05) or social connections do not provide advantages in a meritocracy. Additionally, meritocracies are supported by a cultural belief that all individuals have an equal opportunity to advance in the social class hierarchy.

While meritocracy is often considered an ideal system for social mobility, there are very few truly meritocratic systems. For example, the U.S. class system is not a meritocracy because an individual's SES is partly earned through effort/ability and partly determined by the social class into which they are born.

Lesson 45.3
Spatial Inequality

45.3.01 Residential Segregation

Systems of social stratification impact how societies utilize physical spaces. **Spatial inequality** refers to the uneven and unfair distribution of wealth and resources across a geographic area. Where one resides impacts life opportunities and outcomes. For example, an individual living in a rural area would have access to fewer options in terms of jobs, schools, hospitals, and grocery stores than an individual living in a large city. These differences in the geographic distribution of resources also influence one's lifestyle, social networks (Lesson 36.5), and opportunities for social mobility (Lesson 45.2).

One aspect of spatial inequality is **residential segregation** (ie, separation of social groups into different neighborhoods). As a result, neighborhoods typically include individuals who share common demographic characteristics such as race, class, and ethnicity.

For example, residential segregation occurs when there is a geographical separation between low-income (ie, public housing), middle-income, and wealthy neighborhoods, as depicted in Figure 45.3. Residential segregation is supported by formal laws (eg, historically, some housing loans were regulated based on applicants' race) and informal practices (eg, houses of similar value are built in the same area).

Residential segregation often occurs when neighborhoods are built separately for each income level, which contributes to spatial inequality.

Figure 45.3 Residential segregation and spatial inequality.

45.3.02 Neighborhood Safety and Violence

Spatial inequality and residential segregation impact **neighborhood safety** and **violence**. For example, individuals and families living in low-income neighborhoods (ie, high-poverty neighborhoods) typically experience higher rates of crime and violence than in middle-income or wealthy neighborhoods.

45.3.03 Environmental Justice

Another aspect of spatial inequality includes how elements of the physical environment (eg, parks, landfills, refineries) are managed within a geographic area. Sociologists use the term **environmental justice** to describe environmental policies (eg, air pollution regulations) and practices (eg, building clean water systems) that offer equal protection for all individuals.

However, individuals who are part of marginalized groups often experience a greater burden of environmental hazards. **Environmental injustice** refers to unequal protection from environmental risks and unequal access to environmental benefits within a geographic area. For example, lower-income areas are often subjected to more environmental risk factors such as pollution, toxic waste, and high-voltage power lines, whereas middle-income and wealthy areas have access to more environmental benefits such as paved walkways, safe parks, and clean air.

Lesson 45.4
Poverty

45.4.01 Absolute and Relative Poverty

One consequence of social stratification is **poverty,** which refers to a lack of economic resources (ie, the lower levels of income and fewer assets that are associated with a lower socioeconomic status). Poverty can occur at various levels of society, including individuals, families, communities, and nations.

Sociologists distinguish between absolute and relative poverty (see Figure 45.4), which have different impacts on life outcomes. **Absolute poverty**, a relatively standardized definition, is the inability to secure the basic necessities of life (eg, food, clean water, safe shelter). In the United States, the **poverty line** is used to identify the income level where basic needs cannot be met; the poverty line determines who qualifies for social services (eg, food stamps, subsidized housing).

Alternatively, **relative poverty**, a comparative definition, is the inability to meet the living standards of the society in which one lives. In other words, individuals living in relative poverty can meet their basic needs, but their standard of living is below average. For example, if an individual has permanent housing, but the house or apartment is run-down with broken windows and/or lead paint on the walls, the individual is experiencing relative poverty.

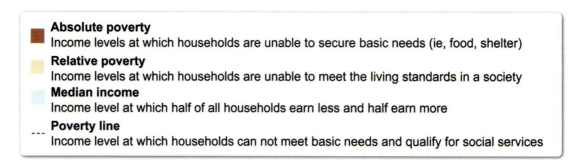

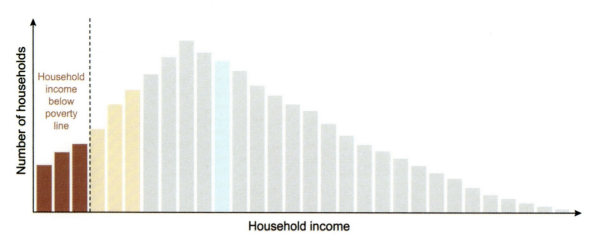

Figure 45.4 Income distribution in a society including absolute and relative poverty.

45.4.02 Social Exclusion

In addition to a lack of material resources, poverty also impacts an individual's social experiences. **Social exclusion** is the marginalization of individuals based on stigmatized identities (eg, mental illness) and disadvantaged situations (eg, poverty), limiting full participation in social life.

In relation to poverty, social exclusion causes individuals to be disconnected and unable to engage in common practices such as earning money through a job. At times, those living in poverty are removed (ie, physically and socially separated) from mainstream society such as when unhoused individuals are removed from city parks by law enforcement.

Lesson 45.5
Health and Healthcare Disparities

45.5.01 Health Disparities

Social stratification affects the delivery of healthcare and individual illness experiences. **Health disparities** describe the inequalities in health outcomes (eg, rates of illness or death) present within society. These patterns of inequality often have a greater impact on disadvantaged groups (eg, racial minorities, individuals in the lower class) as compared to more advantaged groups (eg, individuals who are White, individuals in the upper class).

Sociologists view an individual's health as shaped by elements of society (eg, social institutions) and individual health behaviors. Several macro-level factors such as the physical environment, employment opportunities, and educational systems, as well as micro-level factors such as individual lifestyle choices (eg, diet, exercise), contribute to an individual's health outcomes (see Figure 45.5).

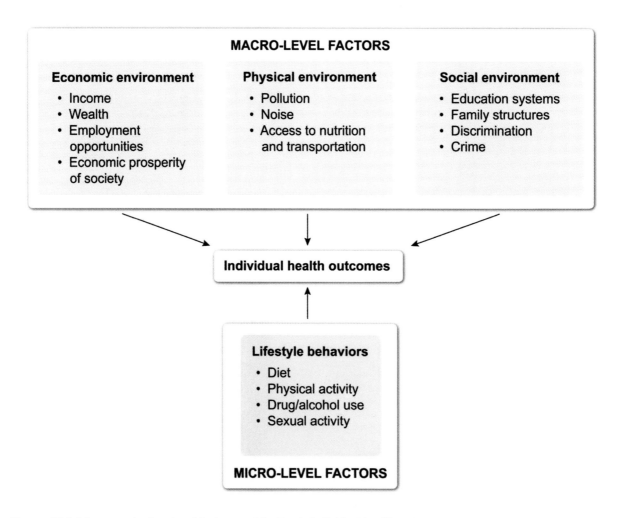

Figure 45.5 Macro- and micro-level factors contributing to individual health outcomes.

The demographic categories of class, gender, and race are also associated with disparities in health outcomes. Middle-class and wealthy individuals are often protected from health risks within the physical environment, such as pollution, waste, and contaminated water (Lesson 45.3). There is also a consistent gender disparity in life expectancy. For example, in the United States, women live an average of five years longer than men. Lastly, racial minorities often experience higher rates of certain illnesses than people who are White, such as people who are Black having the highest rate of hypertension (ie, high blood pressure) compared to all other racial categories.

45.5.02 Healthcare Disparities

Healthcare disparities refer to the way elements of healthcare systems (eg, hospitals, research centers, private insurance versus publicly funded models) contribute to inequality, resulting in unequal access to healthcare services. Many factors contribute to healthcare disparities, including relative poverty (Concept 45.4.01), accessibility issues (eg, language barriers, proximity to clinics/hospitals), and institutional discrimination (Concept 40.3.02) within the healthcare system.

As with health disparities, individuals in marginalized demographic groups often experience greater healthcare disparities. In relation to social class, wealthy individuals typically have greater access to specialized care, whereas those with low incomes often cannot afford expensive treatments, limiting access to needed care.

Another area of healthcare disparity is within medical research used to develop new medications and procedures. Because men have historically been the subjects of medical research studies, a gender disparity in healthcare emerges when treatment protocols for all patients are developed based on data from men only, without considering the possibility of different effects on women's bodies.

There are also racial disparities in healthcare. For example, a study found that some medical students and residents believed Black patients have less sensitive nerve endings (ie, a higher pain tolerance) than White patients. This racial stereotype resulted in Black patients' pain being evaluated as less severe, and thus clinicians were less likely to recommend pain treatment.

Lesson 46.1
Urbanization

46.1.01 Urbanization

One way societies can change is through the process of **urbanization**, which refers to the shift of individuals primarily living in rural, agricultural communities to centralized cities. Another process that influences *urban growth* is **industrialization** (ie, the development of systems of production based on efficiency). When commerce (ie, business and/or trade) and factories expand within cities, the need for individuals to work and live in urban areas increases.

As urban cities grow in both population and commerce, middle-class and upper-class families often leave busy, densely populated city centers and move to the suburbs (ie, neighborhoods developed adjacent to urban areas) for more space and fewer people. **Suburbanization** occurs when there is a large population shift from cities to suburbs; this change can contribute to the *decline of urban areas* because wealth is removed from the cities and transferred to the suburbs.

Often when cities reach a critical level of decline, leaders and businesses attempt to *renew urban areas* through rebuilding city infrastructure (eg, roads, public parks) and renovating buildings. **Gentrification** is a process of urban renewal in which lower-income neighborhoods are revitalized (eg, homes restored, new businesses established) by new, higher-income residents. The influx of economic capital into these neighborhoods results in increased property values. Many long-term residents are then displaced from their neighborhoods due to increases in housing and rental prices.

Lesson 46.2
Globalization

46.2.01 Globalization

In the world today, nations are not independent societies; rather, nations are interconnected with one another. **Globalization** is the process of integrating various cultures and social institutions (eg, economy, government) from different societies by increasing contact and interdependence across the globe. Two of the primary drivers of globalization are advances in communication technology (eg, Internet, cell phones) and a transnational (ie, globally interdependent) economy (eg, global supply chains, international corporations), which result in global rather than local interactions and transactions.

The process of globalization is illustrated in **world systems theory** (see Figure 46.1), which views the world as a global economy where some countries benefit at the expense of others. Within world systems theory, countries are divided into three categories, with each serving a different role in the global economy.

- **Core nations** (eg, United States, United Kingdom) are wealthy countries with diversified economies and strong, centralized governments. Core nations *rely on resources* (ie, raw materials needed to produce goods) from poorer countries and dominate the global economic market through the production and *export of goods* (eg, electronics) around the world.
- **Periphery nations** (eg, Bolivia, Kenya) are developing countries typically with weak governments, limited diversity in the economy, and high levels of inequality. Periphery nations rely on the *export of their resources* (eg, natural gas, coffee) to wealthier countries, making them dependent on and often exploited by core nations.
- **Semi-periphery nations** (eg, India, Brazil) are between core and periphery nations, with economies that are relatively more diversified than those of periphery nations. Semi-periphery nations *export resources* to core nations and produce and *export goods* around the world.

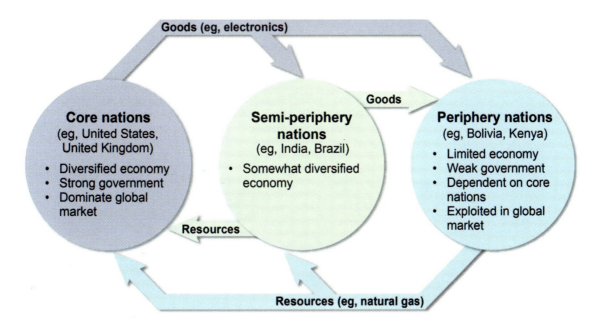

Figure 46.1 Globalization illustrated in world systems theory.

Lesson 46.3
Social Movements

46.3.01 Social Movements

Another way society can change is through actions taken by individuals and/or groups (eg, voting, boycotting). **Social movements** are organized groups that act to either support or reject change in society. Social movements are typically sustained over long periods of time (often for years) through collective actions by individuals who share values and common goals. Notable social movements in the United States include the Civil Rights Movement, various anti-war movements (eg, against the Vietnam War or the Iraq War), and environmental movements (eg, supporting conservation of nature or recycling programs).

Social movement **strategies** refer to how groups mobilize (ie, gather resources such as people and money) and plan ways to promote the movement's goals. Social movement **tactics** are the specific actions taken by those involved in the social movement, such as protests, strikes, and marches.

Typically, the **organization of social movements** (see Figure 46.2) starts with the emergence of an idea (eg, the economic system should have equal opportunities for all people) around which individuals coalesce or unite. Over time, social movements become more organized and bureaucratic and eventually succeed (ie, become mainstream) or fail (ie, dissolve). Social movements may create lasting change in society whether the movement is a success or a failure. For example, in 2011, the Occupy Wall Street movement (ie, a protest against corporate greed) achieved no specific goals, but it brought lasting attention to economic inequality in the United States.

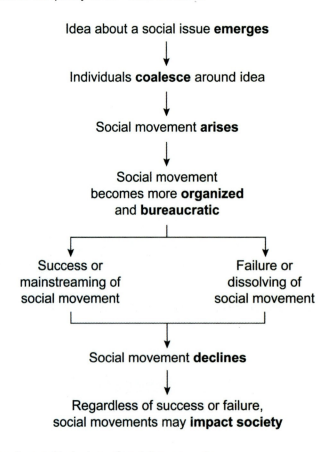

Figure 46.2 Common formation and trajectory of social movements.

Lesson 46.4

Demographic Change

46.4.01 Theories of Demographic Change

Another way societies change is through population growth or decline. In sociology, *demographic shifts* refer to changes in population characteristics (eg, total population, average lifespan) within a society over time, and there are several theories that explain the process of demographic change.

The **demographic transition model** (illustrated in Figure 46.3) refers to demographic shifts in a society from high birth and death rates (as defined in Concept 46.4.03) to low birth and death rates due to changes in the economy (eg, from an agricultural to industrial system) and advances in technology (eg, reproductive medicine). This transition typically occurs in predictable stages:

- Stage 1: In preindustrial societies, birth and death rates are both high and population growth is slow.
- Stage 2: As societies begin to industrialize, death rates drop as food/medicine availability and sanitation increase. Population growth becomes rapid.
- Stage 3: As societies urbanize, the population continues to grow, but birth rates begin to decline as access to contraception increases.
- Stage 4: In developed societies, birth and death rates are both low and population growth is slow, creating a stable population.
- Stage 5: For highly developed societies with very low birth rates, the population may decline; however, Stage 5 is mostly hypothetical because few societies have reached this stage.

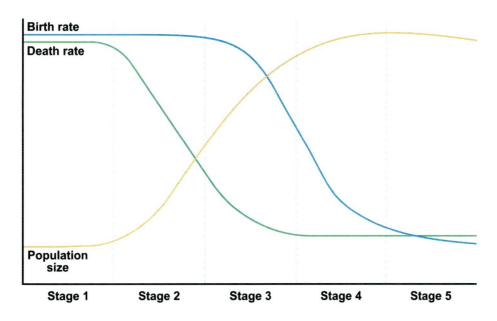

Figure 46.3 Graphed stages of the demographic transition model.

The **Malthusian theory of population growth** suggests that the human population increases exponentially while resources (eg, food) increase linearly at a slower rate. According to this theory, the population growth rate can be slowed by *preventative checks* (ie, a voluntarily decrease in birth rate, such as through the use of contraceptives) and *positive checks* (ie, an involuntary increase in the death rate, which slows or halts overpopulation).

Positive checks can occur through small-scale events such as an increased death rate due to a flu virus within a particular geographic region. Additionally, large-scale positive checks (eg, widespread famine, disease epidemics, wars) dramatically reduce the population by slowing or stopping population growth; as a result, the available resources can more easily sustain the global population.

46.4.02 Population Growth and Decline

Sociologists study patterns of population change to make projections about the future. **Population pyramids** are graphs representing the demographics of a society that provide insights into how the population changes. The graphs display the relative number of males and females by age cohort (Concept 44.1.02) within a population. There are three types of population pyramid shapes: expanding, stationary, and contracting (see Figure 46.4).

Expanding pyramids have broad bases (ie, many younger individuals) and narrow tops (ie, fewer older individuals) and are characteristic of developing countries with high birth and death rates, reflecting an *increasing* population size. **Stationary** pyramids have broad bases and broad tops and are characteristic of developed countries with low birth and death rates and a *stable* population size. **Contracting** pyramids have narrower bases than middles and are characteristic of developed countries with low birth rates and a gradually *declining* population size.

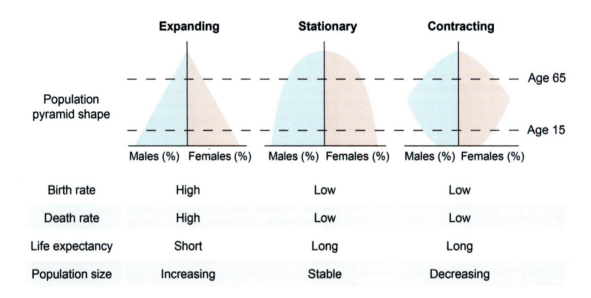

Figure 46.4 Examples and characteristics of the three types of population pyramids.

46.4.03 Fertility and Mortality

Population change is impacted by immigration/emigration rates (Concept 46.4.04), **fertility rates** (also known as birth rates), and **mortality rates** (also known as death rates). Fertility rates measure population increases due to births, whereas mortality rates measure population declines due to deaths.

In sociology, there are multiple ways to report fertility rates:

- **Total fertility rate** (TFR) is the average number of children born per woman during her lifetime. A TFR above 2 (referred to as the replacement rate) means the population is growing, and a TFR below 2 means the population is shrinking.
- **Crude birth rate** (CBR) is the number of live births per year for every 1,000 members of a population, which is a rough estimate of annual population growth based on birth only. For example, Uganda (ie, a developing nation) has a higher CBR (around 40), whereas Japan (ie, a developed nation) has a lower CBR (around 7).

- **Age-specific fertility rate** (ASFR) is the number of live births per year for 1,000 women in a certain age group in a population. For example, in 2019 in the United States, the ASFR for women ages 25–29 is about 94, whereas the ASFR for women ages 40–44 is about 13.

The mortality rate (ie, the number of people who die within a population during a specific time period) also impacts population change. Like fertility rates, mortality rates can be reported in several ways. The **crude death rate** refers to the number of deaths per year for every 1,000 members of a population. **Age-specific mortality rates** measure the rate of death within certain age cohorts (Concept 44.1.02) in a population. One example of an age-specific mortality rate is the *infant mortality rate*, which measures the number of deaths of individuals under one year of age per 1,000 live births in a year.

46.4.04 Push and Pull Factors in Migration

In sociology, another factor in population change is **migration**, which refers to the movement of individuals within a society (see Lesson 44.5) through emigration (ie, relocation out of an area) and immigration (ie, relocation into an area). Migration patterns can be explained by push and pull factors impacting emigration and immigration, respectively. **Push factors** (eg, natural disasters, war) describe why people move away from their country of origin, whereas **pull factors** (eg, education opportunities for women, economic prosperity) describe why people move to a new country.

END-OF-UNIT MCAT PRACTICE

Congratulations on completing **Unit 9: Demographics and Social Structure**.

Now you are ready to dive into MCAT-level practice tests. At UWorld, we believe students will be fully prepared to ace the MCAT when they practice with high-quality questions in a realistic testing environment.

The UWorld Qbank will test you on questions that are fully representative of the AAMC MCAT syllabus. In addition, our MCAT-like questions are accompanied by in-depth explanations with exceptional visual aids that will help you better retain difficult MCAT concepts.

TO START YOUR MCAT PRACTICE, PROCEED AS FOLLOWS:

1) Sign up to purchase the UWorld MCAT Qbank
   IMPORTANT: You already have access if you purchased a bundled subscription.
2) Log in to your UWorld MCAT account
3) Access the MCAT Qbank section
4) Select this unit in the Qbank
5) Create a custom practice test

Index

A

absolute refractory period, 23
absolute threshold, 49
accommodation (in vision), 58
accommodation (Piaget's theory of cognitive development), 122
acetylcholine, 27
acquisition, 91
action potential (AP), 19–26, 42, 118
activation-synthesis hypothesis, 76, 79
actor-observer bias, 218–19
adrenal glands, 18, 38, 157
adrenocorticotropic hormone, 38, 157
age cohorts, 247, 268–69
aggression, 3, 29, 104, 143, 213
agonists, 28, 167–68
agoraphobia, 163
alertness, 75
algorithms, 125
altruism, 215
amnesia, 29, 115, 165
amygdala, 29, 51, 69, 83, 143, 158, 213
anomie, 233
antagonists, 28
anxiety disorders, 163
appraisal theory, 155
archetypes, 149
arousal theory, 137
assimilation (into a culture), 191–92
assimilation (Piaget's theory of cognitive development), 121
attachment, 214
attention
  divided, 87
  selective, 87
attitudes, 145–46, 159, 194, 223
attraction, 212, 249
attribution theory, 216
auditory cortex, 31, 63
auditory localization, 64
authority, 244
autonomic nervous system, 16–18, 27, 143, 157
availability heuristic, 126

B

basal ganglia, 29–30, 168
basilar membrane, 50, 62–63
behaviorist approach, 4, 129, 169–70
Big Five theory, 153
biological preparedness, 92, 102
biomedical approach, 169
biopsychosocial approach, 169
bipolar disorder, 164
bottom-up processing, 52
brainstem, 30, 67, 74–75
Broca's area, 32, 132
bureaucracy, 209–10, 242
  ideal type, 209–10
bystander effect, 225–26

C

capital, 208, 253
central nervous system (CNS), 15–17, 26, 34, 81–82
cerebellum, 30, 74
chemical senses, 51, 67
chemoreceptors, 50–51, 67
chunking, 110
churches, 242–43
circadian rhythms, 37, 79
class consciousness, 253
classical conditioning, 90–94, 100, 102, 163
cochlea, 62
cocktail party effect, 87
cognitive dissonance, 147
cognitive maps, 100–101
collective unconscious, 149
color perception, 60
computerized tomography, 42, 245
conditioned response, 90–92
conditioned stimulus, 90–92, 98
conditioned taste aversion, 90, 92–93
confirmation bias, 126–27
conflict theory, 176
conformity, 198, 227
consciousness, 30, 75–76, 148, 158, 165
conservation, 120
context-dependent effects, 110
convergence, 54
corpus callosum, 32

correlational studies, 7
correlation coefficient, 11
cortisol, 38, 157
countercultures, 190
cults, 242–43
cultural diffusion, 193
cultural lag, 191
cultural relativism, 221–22
cultural transmission, 193
culture
    components of, 187–90, 193
    material, 189, 191
    popular, 192
    symbolic, 189, 191
culture shock, 191

D

decision-making, 31, 125–28, 165, 177, 229–30
defense mechanisms, 148
deindividuation, 226
demographics, 175, 247, 251, 267–68
dependence, 81–82
    physical, 82
    psychological, 82
dependent variable, 6–7
depressants, 81
depth cues
    binocular, 54
    monocular, 53–54
depth perception, 53–54
descriptive statistics, 10
desensitization, 88–89
deviance, 234, 245
difference threshold, 47–48
differential association theory, 234
discrimination
    in classical conditioning, 92
    in operant conditioning, 98
    individual, 223
    institutional, 223, 256, 263
dishabituation, 88–89
dissociation, 75
dissociative disorders, 165
division of labor, 210, 244
door-in-the-face strategy, 146
dopamine, 27, 83, 167–69
dramaturgical approach, 200
dreaming, 30, 75–77, 79
drive reduction theory, 138
Durkheim, Émile, 176

E

educational stratification, 240
egocentrism, 120
elaboration, 106–7
elaboration likelihood model, 159
electrical stimulation of the brain, 43
electroencephalography, 42, 77
emigration, 251, 269
emotion
    components of, 140
    theories of, 141
empiricism, 179, 243
encoding, 106, 108–9
environmental justice, 259
epinephrine, 18, 27, 157
Erikson, Erik, 197
ethnicity, 220, 250, 252, 258
ethnocentrism, 221–22
expectancy, 100
expectancy theory of motivation, 139
experiment, 6
experimental study, 7
extinction, 91–92, 96, 98

F

false consciousness, 253
family
    abuse in, 241
    forms, 241
feature detectors, 59
feminist theory, 177–78
fertility rates, 268–69
fetal alcohol syndrome, 41
fight-or-flight, 18, 157
folkways, 231
foot-in-the-door strategy, 146
forgetting curve, 113
fovea, 58
frequency theory, 63–64
Freud, Sigmund, 3, 75, 79, 148
frontal lobe, 31, 105
functional fixedness, 128
functionalism, 176
functional magnetic resonance imaging, 42
fundamental attribution error, 217–18
fundamentalism, 243

G

gamma-aminobutyric acid (GABA), 26–27, 81
gate-control theory, 66
gender, 177, 196, 247–48, 252, 254, 262

gender roles, 178, 248
gender segregation, 248
general adaptation syndrome, 157–58
generalization
    in classical conditioning, 92
    in operant conditioning, 98
generalized anxiety disorder, 163
gentrification, 264
Gestalt principles, 53, 55
glial cells, 15, 19–20
globalization, 265
glutamate, 26, 67
Goffman, Erving, 200
government, 209, 244
group
    dyad, 206
    primary, 205
    reference, 206
    secondary, 205
    social, 177, 180, 196, 198, 205, 221, 223, 232, 258
    triad, 206
group polarization, 229
groupthink, 230
gustation, 67

H
habituation, 88–89
hair cells, 50–51, 62, 73–74
hallucinogens, 81
healthcare disparities, 262–63
health disparities, 262–63
hemispheric lateralization, 32, 66
hidden curriculum, 239
hierarchy of needs, 139, 151
hindsight bias, 127
hippocampus, 29, 31, 51, 69, 117, 158
hormones, 29, 36–38, 79, 143, 169, 245
humanistic approach, 5, 151, 170
hypothalamus, 29, 36–38, 79, 143–44, 157
hypothesis
    alternative, 6
    null, 6

I
identity
    gender, 178, 248
    illness and, 245
    race and, 250
    religion and, 242
    sexual orientation and, 249
    social, 196, 198, 202, 205, 220, 252, 254

identity formation, 197–99
illness experience, 245–46
immigration, 191, 251, 269
impression management, 201
incentive theory, 138–39
independent variable, 6–7
industrialization, 243, 264
inferential statistics, 11
informed consent, 9
in-group, 205, 220
insomnia, 80
instinct, 102–3, 137
intelligence, 123–24
interference
    proactive, 113–14
    retroactive, 113–14
intersectionality, 254–55
iron law of oligarchy, 210

J
just-noticeable difference, 47
just-world hypothesis, 219

K
kinesthetic sense, 71, 74
kinship, 241
Kohlberg, Lawrence, 197–98

L
labeling theory, 234
language, 31, 110, 120, 129–32, 187, 194–95, 199, 220, 250
language acquisition device, 129
learning
    associative, 90, 95, 100, 102, 116
    avoidance, 99
    escape, 99
    non-associative, 88–90, 116
    observational, 104–5, 213
    social, *see* observational
lesioning, 43
levels of processing, 106
life course approach, 247
limbic system, 29, 143
linguistic relativity hypothesis, 131
long-term potentiation, 116–18
looking-glass self, 199

M
macrosociology, 175
magnetic resonance imaging, 42, 245
major depressive disorder, 164–65, 167

Malthusian theory of population growth, 267
Marx, Karl, 176, 253
Maslow, Abraham, 5, 139, 151–52
mass media, 192, 194
McDonaldization, 210–11
Mead, George Herbert, 177, 199
mechanoreceptors, 50–51, 62, 65, 71
medicalization, 245
meditation, 158
melatonin, 29, 79
membrane potential, 21–22
memory
   explicit, 110
   implicit, 110
   long-term, 109–10, 113, 117–18
   sensory, 109
   short-term, 109–10, 113, 118
memory reconstruction, 114
mental set, 128
meritocracy, 187, 257
mesolimbic reward pathway, 83
microsociology, 175
midbrain, 29–30, 59, 63, 83, 168
mirror neurons, 105
misinformation effect, 114
mnemonics, 108–9
modernization, 243
moral development, 197–98
mores, 231
mortality rates, 268–69
motivation, 29, 83, 137–39, 147
motivational conflict theory, 156
motor cortex, 31, 43, 105
motor skills, 41, 110, 115
multiculturalism, 191–92

N
narcolepsy, 80
networks, 207–8
neural plasticity, 116
neuroimaging, 42–43, 158
neurotransmitters, 19–21, 24–28, 81–82, 116–18, 167, 169. *See also* acetylcholine; dopamine; epinephrine; gamma-aminobutyric acid (GABA); glutamate; norepinephrine; serotonin
neutral stimulus, 90–93, 100
night terrors, 80
nociceptors, 65–66
non-normative behavior, 231–35, 245
norepinephrine, 18, 27, 157, 167
normative behavior, 231–32, 234
norms, 191, 200, 224, 231–34, 248, 254
nucleus accumbens, 30, 83

O
obedience, 198, 227–28
object permanence, 119
obsessive-compulsive disorder, 163–64
occipital lobe, 31, 58–59, 116
olfaction, 29, 51, 67, 69–70
olfactory bulb, 29, 51, 69
olfactory receptor neurons, 69
operant conditioning, 95, 97–99, 102–3, 129–30
opioids, 27, 81
opponent process theory, 60
optic nerve, 58–59
organizations
   formal, 209, 242
   informal, 209, 242
ossicles, 61
otolith organs, 62, 73–74
out-group, 205, 220

P
panic attacks, 163
panic disorder, 163
parallel processing, 59
parasympathetic nervous system, 17, 27
parietal lobe, 31, 60, 66, 71
Parkinson's disease, 27, 30, 166, 168
Pavlov, Ivan, 4, 90
perceptual constancy, 53
perceptual organization, 53, 55
perceptual set, 52
peripheral nervous system (PNS), 15–17, 34
personality, 31, 148–53, 196
phi phenomenon, 55
photoreceptors, 50–51, 57–58, 79
Piaget's theory of cognitive development, 119, 121
pineal gland, 29, 79
pitch perception, 63–64
pituitary gland, 36–38, 143, 157
placebo, 6, 8. *See also* placebo effect
placebo effect, 8
place theory, 63–64
pons, 30, 76, 79
population pyramids, 268
positron emission tomography, 42
posttraumatic stress disorder, 163–64, 245
poverty, 253, 260–61

power, 137, 176–78, 202, 210, 220, 223–24, 243–44, 252–54, 256
prefrontal cortex, 30–31, 75, 83, 158
prejudice, 196, 220–23
prestige, 178, 220, 223, 254, 257
privilege, 176, 254
problem-solving, 125–26, 128, 220
proprioception, 31, 66, 71
proprioceptors, 50, 71, 74
psychoanalytic approach, 3, 148
psychophysics, 47
punishment, 95, 97
push and pull factors, 269

R

race, 196, 202, 205, 220, 240, 247, 250, 252, 254, 258, 262
racial formation theory, 250
racialization, 250
rational choice theory, 177
reflexes, 35, 39, 41, 102, 110, 137
reinforcement, 95–97, 100, 103, 129–30
  schedules of, 95–97
reinforcer
  conditional, 98
  primary, 98
  secondary, 98
reliability, 8
religion, 189, 239, 241–43
religiosity, 242
REM sleep, 77–79
representativeness heuristic, 126
residential segregation, 258
reticular activating system, 75
reticular formation, 30, 75
retina, 51, 54, 57–60
retinal disparity, 54
retrieval, 106, 108–10, 114
rituals, 187, 190, 192
role conflict, 204
role exit, 204
role performances, 200, 203
role strain, 204
roles, social, 145, 199–200, 203, 245, 247

S

salience, 196
sanctions, 232
Sapir-Whorf hypothesis. *See* linguistic relativity hypothesis
schemas, 121–22, 220
schizophrenia, 27, 165, 167, 169
sects, 242–43
secularization, 243
self-actualization, 151–52
self-concept, 152, 196, 206
self-fulfilling prophecy, 221, 239
self-reference effect, 107–8
self-serving bias, 216–17
semicircular canals, 62, 73–74
sensitization, 88–89
sensory adaptation, 47
sensory interaction, 67, 74
serial position effect, 111
serotonin, 27, 167
sex, 202, 248–49
sexual orientation, 196, 247, 249
shaping, 97–98
sick role theory, 245
signal detection theory, 49
Skinner, B.F., 4, 95, 129
sleep apnea, 80
sleep cycles, 77–78
sleep stages, 42, 77–78
sleep-wake disorders, 78, 80
social anxiety disorder, 163
social class, 252, 254, 257, 263
social constructionism, 177
social constructs, 177, 234, 250
social control, 224, 232, 250
social epidemiology, 246
social exchange theory, 177
social exclusion, 261
social facilitation, 224
social institutions, 176, 192, 194, 239, 241–42, 245, 262, 265
social loafing, 225
social mobility, 256–58
social movements, 176, 266
social reproduction, 254
social stratification, 252, 254, 260, 262
socialization, 193–95, 198, 231, 234, 239
  agents of, 194, 198
  primary, 195
  secondary, 195
socioeconomic gradient in health, 255
socioeconomic status, 180, 252, 255, 257, 260
somatic nervous system, 16–17, 27
somatosensation, 31, 65–66
somatosensory cortex, 31, 65–66, 71
somnambulism, 80
source monitoring errors, 114
spatial inequality, 258–59

specific phobia, 94, 102, 163, 169
spinal cord, 15–17, 27, 30, 34–35, 65–66, 71
spinal reflex, 34–35, 66
spontaneous recovery, 91–92
Stanford prison experiment, 145
state-dependent effects, 110
status, 145, 189, 202–3, 248
status set, 202
stereotypes, 220–23, 248, 263. *See also* stereotype threat
stereotype threat, 221
stigma, 221
stimulants, 81
storage, 106, 109–10, 114
strain theory, 234
stressors, 17, 38, 155–57
subcultures, 190
substance use disorders, 83
substantia nigra, 30, 166, 168
suburbanization, 264
symbolic interactionism, 176–77
symbols, 176–77, 187, 189–92, 199
sympathetic nervous system, 17–18
synapse, 19–20, 25–26, 36, 59, 82, 116–18, 167
systematic desensitization, 169–70

T
taboos, 231
taste buds, 51, 67
teacher expectancy, 239
temporal lobe, 31, 60, 63, 132
teratogen, 41
thalamus, 29, 51, 58–59, 63, 65–67, 70–71, 74
thermoreceptors, 50–51, 65
tip-of-the-tongue phenomenon, 112
tolerance, 82–83
top-down processing, 52
trait theory, 153
transduction, 47, 50, 65, 67
trichromatic theory, 60

U
unconditioned response, 90, 92
unconditioned stimulus, 90–93, 100
unconscious mind, 3, 76, 148
universal emotion theory, 140
urbanization, 264

V
validity, 8
ventral tegmental area, 30, 83
vestibular sense, 73
visual cortex, 31, 51, 58–59

W
Weber, Max, 209, 244
Weber's law, 48
Wernicke's area, 31–32, 129, 132
withdrawal, 82–83
world systems theory, 265